ALL THINGS BEING EQUAL

All Things Being Equal

Why Math Is the Key to a Better World

JOHN MIGHTON

ALFRED A. KNOPF CANADA

PUBLISHED BY ALFRED A. KNOPF CANADA

 Published in 2020 by Alfred A. Knopf Canada, a division of Penguin Random House Canada Limited, Toronto. Distributed in Canada by Penguin Random House Canada Limited, Toronto.

www.penguinrandomhouse.ca

Library and Archives Canada Cataloguing in Publication

Title: All things being equal : why math is the key to a better world / John Mighton.
Names: Mighton, John, 1957- author.
Identifiers: Canadiana (print) 20190158557 | Canadiana (ebook) 20190158611 |
ISBN 9780735272897 (hardcover) | ISBN 9780735272910 (HTML)
Subjects: LCSH: Mathematics—Social aspects. | LCSH: Numeracy.
Classification: LCC QA10.7 M54 2020 | DDC 510—dc23

Text design by Five Seventeen
Cover design by Five Seventeen
Image credits: © Audrey Shtecinjo / Stocksy

Printed and bound in Canada

10 9 8 7 6 5 4 3

For Pamela Mala Sinha

CONTENTS

INTRODUCTION

Nothing comes easily to me.

I'm a mathematician, but I didn't show much aptitude for math until I was thirty. I had no idea, in high school, why I had to turn a fraction upside down when I wanted to divide by it, or why, when I wrote a square root sign over a negative number, the number suddenly became "imaginary" (especially when I could see the number was still there). At university I almost failed my first calculus course. Fortunately I was saved by the bell curve, which brought my original mark up to a C minus.

I'm also a playwright. My plays have been performed in many countries, but I still won't read a review unless someone tells me it's safe to do so. Early in my career I made the mistake of checking the papers to see what two of the local critics thought of my first major production. It seems unlikely that they consulted each other before writing their reviews, but one headline read "Hopelessly Muddled" and the other "Muddled Mess."

I often wish I was more like my literary and scientific heroes, who seemingly could produce perfect poems or solve intractable problems in a blinding flash of inspiration. Now that I'm a professional mathematician and writer, I console myself with the thought that my ongoing struggles to educate myself and the strenuous efforts that I needed to make to get to this point have produced an intense curiosity about how we achieve our potential.

A Slow Learner

From an early age I became obsessed with my intellectual capabilities and with the way I learn. When I started to teach in my twenties, first as a graduate student in philosophy and later as a math tutor, I also became fascinated with the way other people learn. Now, after teaching math and other subjects to thousands of students of all ages and after reading a great deal of educational and psychological research, I am convinced that our society vastly underestimates the intellectual potential of children and adults.

During my undergraduate studies, I showed as little promise in writing as I did in mathematics: I received a B plus in my creative writing class—the lowest mark in the class. One evening, in the first year of my graduate studies in philosophy, I began reading a book of letters by the poet Sylvia Plath, which I'd found on my sister's bookshelf while babysitting her children. It appeared from Plath's letters and early poems that she had taught herself to write by sheer determination. She had learned, as a teenager, everything she could about poetic metre and form. She wrote sonnets and sestinas, memorized the thesaurus and read mythology. She also produced dozens of imitations of poems she loved.

I knew that Plath was considered to be one of the most original poets of her time, so I was surprised to learn that she had taught herself to write by a process that seemed so mechanical and uninspired. I'd grown up thinking that if a person was born to be a writer or mathematician, then fully formed and profoundly important sentences or equations would simply pour out of them. I'd spent many hours sitting in front of blank pages waiting for something interesting to appear, but nothing ever did. After reading Plath's letters, I began to hope that there might be a path I could follow to develop a voice of my own.

I imitated the work of Plath and other poets for several years before I moved on to writing plays. By that time, I'd taken a job at a tutoring agency to supplement my income from writing. The women who owned the agency hired me to tutor math because I'd taken a course in calculus at university (and I neglected to tell them about my marks). In my tutorials I had the opportunity to work through the same topics and problems again and again with my students, who ranged in age from six to sixteen. The concepts that had mystified me as a teenager (such as why does a negative times a negative equal a positive) gradually became clear, and my confidence grew as I found I could learn new material more quickly.

One of my first students was a shy eleven-year-old boy named Andrew, who struggled in math. In grade six, Andrew was placed in a remedial class. His new teacher warned his mother that she shouldn't expect much from her son because he was too intellectually challenged to learn math in a regular math class. In the first two years of our tutorials, Andrew's confidence grew steadily, and by grade eight, he had transferred to the academic stream in math. I tutored him until he was in grade twelve, but I lost touch with him until recently when he invited me to lunch. In the middle of our lunch, Andrew told me that he had just been granted full tenure as a professor of mathematics.

When I was growing up I would always compare myself to the students who did well on math competitions and who seemed to learn new concepts without effort. Watching these students race ahead of me at school made me think that I lacked the natural gift I needed to be good at the subject. But now, at the age of thirty, I was surprised to see how quickly I could learn the concepts I was teaching, and how easy it was for students like Andrew, who had never shown any signs of having a "gift" for math, to excel at the subject with patient teaching. I began to suspect that a root cause of many individuals' troubles in math, and in other subjects as well, is the belief in natural talents and natural academic hierarchies.

As early as kindergarten, children start to compare themselves to their peers and to identify some as talented or "smart" in various subjects. Children who decide that they are not talented will often stop paying attention or making an effort to do well (as I did in school). This problem is likely to compound itself more quickly in math than in other subjects, because when you miss a step in math it is usually impossible to understand what comes next. The cycle is vicious: the more a person fails, the more their negative view of their abilities is reinforced, and the less efficiently they learn. I will argue that the belief in natural hierarchies is far more instrumental in causing people's different levels of success in math and other subjects than are inborn or natural abilities.

In my early thirties, I returned to university to study mathematics (starting at an undergraduate level) and was eventually granted one of Canada's highest postdoctoral fellowships for my research in the subject. In the meantime, I'd also received several national literary awards for my plays, including a Governor General's Award. I don't believe I will ever produce work that compares to that of my artistic and intellectual heroes, but my experience suggests that the methods I used to train myself as a writer and mathematician—which included deliberate practice, imitation, and various strategies for mastering complex concepts and enhancing the imagination—could help people improve their abilities in the arts and sciences.

When I was taking my degree in math, I often wondered how my life would have gone if I'd selected a different book from my sister's bookshelf the night I discovered Plath's letters. I felt lucky to have regained the passion I'd had as a child for creating and discovering new things and lucky to have been encouraged to follow my passion by my parents and family. Watching my students become more engaged and successful in math, I began to feel that I should do something to help people who'd lost faith in their abilities, so they could regain their confidence and keep their sense of wonder and curiosity alive.

In the final year of my doctoral program, I persuaded some of my friends to start a free, after-school tutoring program called JUMP (Junior Undiscovered Math Prodigies) Math in my apartment. Twenty years later, 200,000 students and educators in North America use JUMP as their main math instruction resource, and the program is spreading into Europe and South America. Its methodologies have been developed in consultation with and guided by the work of distinguished cognitive scientists, psychologists and educational researchers, many of whom you will meet in this book. These methods are easy to understand and apply, and they reinforce confidence in your abilities rather than assigning you to a particular skill level. They can be used by adults who want to help children learn any subject more efficiently or who want to educate themselves and pursue a new path in life, the way I did.

Before I describe these methods and the research that supports them, I will look more closely at some myths about intelligence and talent that prevent us from fully developing our intellectual abilities and that create an extraordinary range of problems for our society. Because people have so much trouble imagining that they could be good at subjects they struggled to learn in school, they also have trouble imagining what the average brain can accomplish or understanding the magnitude of the losses our society incurs when we fail to educate people according to their potential. This failure of the imagination creates a self-fulfilling cycle of frustration and lost opportunities for many people; to escape from this cycle we need to re-examine our most basic beliefs about what it means for people to be "equal" or to have equal opportunities in life.

Invisible Problems

Every society is plagued by invisible problems that are particularly hard to solve—for no reason other than because they are invisible. Sometimes a society has to collapse before the problems that stopped it from progressing can be seen. And sometimes this process can take centuries.

The ancient Greeks were remarkable innovators. They established the first democracies and produced a staggering number of mathematical and scientific breakthroughs. But this great, progressive society was hobbled by an insidious problem they could not see. Even the most enlightened thinkers of 400 BC were convinced that women were inferior to men and that slavery was as good for slaves as it was for their owners. Aristotle wrote, rather chillingly, that some people are born to be masters, while others are only fit to be "living tools." The Greeks couldn't begin to solve the most serious problems of their time because they couldn't conceive of a more equitable society.

Over the past three hundred years, the idea that every person is born with the same inalienable rights and privileges, regardless of their race, gender or social status, has slowly taken hold across the world. In theory, in most nations, we have all been granted these same rights.

In practice, however, these rights are not always upheld in the same way for every person. And in many parts of the world, the impact of these rights on people's quality of life is still rather limited. Even in Western democracies, people who are born with the same inalienable right to vote don't necessarily enjoy the same social or economic opportunities.

Half of the world's wealth is owned by 1 percent of the population, and tens of millions of people still don't have enough to eat or proper access to health care or sanitation, even in the developed world. We are confronted by an array of threats—including economic instability, climate change, sectarian violence and political corruption—that have a greater impact on the poorest and most disadvantaged people of the world. In such a world, it's hard to imagine a society in which people are born "equal" in any material sense or in which they can exercise their basic legal and political rights in the same way.

The laws and constitutions created to give everyone a fair chance in life have only partially succeeded in levelling the playing field. That's because the most serious disparities in our society are not simply the result of legal or political inequalities but are also caused by a more subtle and pervasive form of inequality that is difficult to see. This kind of inequality might seem to be a by-product of social and political forces or of the deficiencies of capitalism, but I believe it is primarily caused by our ignorance about human potential. In the developed world, this inequality can affect the children of the rich as much as it does the children of the poor (although wealth does help mitigate its consequences). In many ways, it is the root cause of other inequalities. I call this kind of inequality "intellectual inequality." And I will argue that it can easily be eradicated, particularly in the sciences and mathematics.

In this book I will sometimes use examples from JUMP Math to illustrate various principles of learning and teaching. But this is not a book about JUMP. My claims about human potential and the methods of teaching and learning that can unlock that potential are backed by a large body of research in cognitive science and psychology that is independent of JUMP. One day this research will be more widely known, and we will all be compelled to set much higher expectations in mathematics and other subjects for ourselves and our children, whether or not we use any particular math learning program. When we have understood and absorbed the full meaning of this research, our present beliefs about our intellectual abilities will seem as antiquated and as harmful as the belief that some people are born to be slaves and others masters. And the problems we have struggled to overcome since antiquity—which originate in our failure to foster intellectual equality—may finally be addressed.

PART ONE:
WHY MATH?

CHAPTER 1: THE 99 PERCENT SOLUTION

Over the past two decades, research in cognitive science has radically changed the way scientists think about the brain. Researchers have discovered that our brains are plastic and can learn and develop at any stage of life. As well, a growing body of evidence suggests that the vast majority of children are born with the potential to learn anything, particularly if they are taught by methods that have been shown to be effective. A variety of psychological studies—in which people have been trained to develop musical abilities that were once considered to be innate (like perfect pitch) or to significantly improve their performance on SAT tests (by becoming better at seeing analogies)—indicate that experts are made not born. As Philip E. Ross points out in "The Expert Mind", a research survey published in *Scientific American*, these results have profound implications for education. According to Ross, "Instead of perpetually pondering the question, "Why can't Johnny read?" perhaps educators should ask, "Why should there be anything in the world that he can't learn to do?"

Many people believe that math is an inherently difficult subject—accessible only to people who are born with a "gift" with numbers or who display mathematical ability at an early age—but I will present evidence that math is the subject in which learners of all ages can most easily unlock their true intellectual potential. Indeed, if every child was taught according to their true potential from the first day of school, then I would predict that by grade five, 99 percent of students could learn and love learning math as much as the top 1 percent do now. And I believe that the significant majority of adults could develop an aptitude for math if they were taught using the methods that I will demonstrate in this book.

When people are presented with evidence that contradicts their long-held beliefs, they often find ways to ignore that evidence or explain it away. Psychologists call this way of dealing with conflicting viewpoints "cognitive dissonance." As a society, we have for many years been living in a state of extreme cognitive dissonance regarding the human capacity to learn. I remember reading newspaper articles about the remarkable intellectual potential of children and the surprising plasticity of older brains as long ago as the 1990s. Since then I've read many excellent books on this topic, including David Shenk's *The Genius in All of Us* and Carol Dweck's *Mindset*. I wrote about the issue in my own books, *The Myth of Ability* and *The End of Ignorance*.

It strikes me as odd that although the research has long been widely publicized, its existence has done very little to change the way that people think about their own intellectual abilities or the way people are taught—at home, at school or in the workplace. Just as the ancient Greeks couldn't conceive of a world in which every person is born free, it appears that, in spite of the evidence, we can't conceive of a world in which virtually every person is born with the potential to learn and love learning any subject—including difficult seeming subjects like math and science.

Intellectual Hierarchies Make Everyone Less Intelligent

To see the extent to which we live in a state of cognitive dissonance, let us consider an example. When people complain about problems in North American education, they often speak as if those problems would be solved if students in the US and Canada were able to perform as well on international tests of reading and mathematics as students from countries that achieve the highest scores. Nations like Finland and Singapore are singled out in the media as having superior educational systems because their students do better on standardized tests of mathematical achievement like the PISA test, the Programme for International Student Assessment test written by fifteen-year-old students in eighty countries every three years.

It's worth looking at the results of these tests closely, but more for what they reveal about our beliefs about children and their potential than for what they prove about education. From the way people talk about the tests, you can clearly see what they expect the average child to achieve at school.

On the PISA test, a score at level 5 or 6 in mathematics is required to take courses at university; a score at level 3 or below suggests the test taker would have trouble holding a job that required much more than a basic knowledge of math. In 2015, only 6 percent of American students and 15 percent of Canadian students scored at level 5 or 6, compared with 12 percent in Finland and 35 percent in Singapore. However, in Finland almost 55 percent of students scored at level 3 or below and in Singapore about 40 percent of students scored at these levels. (The results for the US and Canada were 79 percent and 62 percent, respectively.)[2]

Many people have suggested that American educators should find out how math is taught in the top-performing countries so it can be taught in the same way in the US. I expect this is a good idea, but we might also want to find out how countries that produce such strong students still manage to teach so little to almost half their populations. Answering this question might do as much to help us improve the teaching of mathematics as any effort to emulate the educational practices of other countries.

Wide differences in mathematical achievement among students appear to be natural. In every school in every country, only a minority of students are expected to excel at or love learning mathematics. In the many schools I have visited, on several continents, I have always seen a significant number of students who are two or three grade levels behind by the end of elementary school. In my home province of Ontario, where children do rather well on international tests, fewer than 50 percent of grade six students met grade-level standards on the 2018 provincial exams. And the same differences can be found in other subjects, particularly in the sciences.

In my work with children and adults, I have seen a great deal of evidence that mathematical ability is extremely fluid and that teachers can produce dramatic improvements in achievement with very simple interventions. The extent to which we underestimate the mathematical abilities of the general population is reflected in a case study involving a Canadian elementary school class. This study was first reported in the *New York Times* and was later featured in *Scientific American Mind*.

In the fall of 2008, a fifth-grade teacher from Toronto, Mary Jane Moreau, tested her students on a standardized test called the TOMA (Test of Mathematical Abilities). This graph shows the distribution of marks in her class.

The average mark for the class was in the 54th percentile, with the lowest mark in the 9th percentile and the highest in the 75th. This range of marks would represent about a three-grade-level difference between the top and bottom of the class. (A fifth of this class were diagnosed with learning disabilities.)

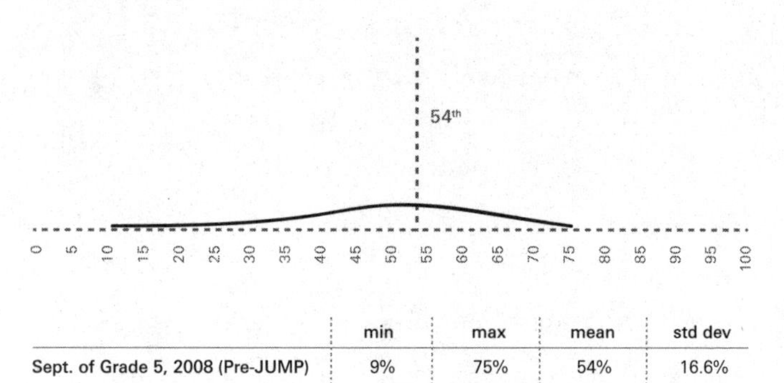

	min	max	mean	std dev
Sept. of Grade 5, 2008 (Pre-JUMP)	9%	75%	54%	16.6%

*Class percentile ranking based on results on the norm-referenced Test of Mathematical Abilities

I've surveyed several hundred grade five teachers in my talks and training sessions, and they've all reported similar differences in their students. These differences grow even more pronounced as students get older. By high school many are "streamed" into applied or essentials courses, while others struggle to keep up in academic courses. Moreau's students were enrolled in a very good private school. Their test results show the extent to which we have come to accept these inequalities as natural. Even the most affluent parents are happy to send their children to schools that produce vastly different results for individual students, which suggests that intellectual inequality is not primarily an economic or class problem.

I met Mary Jane Moreau when she introduced herself after one of my presentations about JUMP Math at a local teachers' college. She was an innovative teacher who had taught at a laboratory school called the Institute of Child Study before transferring to a private school. She had a deep interest in educational research and enjoyed experimenting with new pedagogical approaches, so she decided to investigate the JUMP program for herself. After testing her students on the TOMA, she abandoned her usual approach of pulling together lessons with the best materials she could find and followed the JUMP lesson plans with fidelity. This meant teaching concepts and skills in steps that were much smaller than the steps she normally followed, constantly asking questions and assigning exercises and activities to assess what her students knew, giving frequent practice and review, and most importantly, building excitement by giving students incrementally harder series of challenges where one idea builds on the next. I will expand on this method of "structured inquiry" (in which students discover concepts and figure things out for themselves, but with sufficient and rigorous guidance from the teacher) in chapter four.

After a year of JUMP, Moreau retested her students on the grade six TOMA. The average score of her students rose to the 98th percentile with the lowest mark in the 95th percentile (see graph on page 20).

At the end of grade six, Moreau's entire class signed up for the Pythagoras math competition, a prestigious contest for sixth graders. One of the strongest students was absent on the day of the exam, but of the seventeen

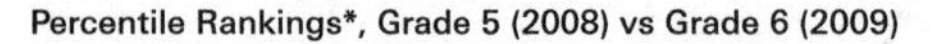

Percentile Rankings*, Grade 5 (2008) vs Grade 6 (2009)

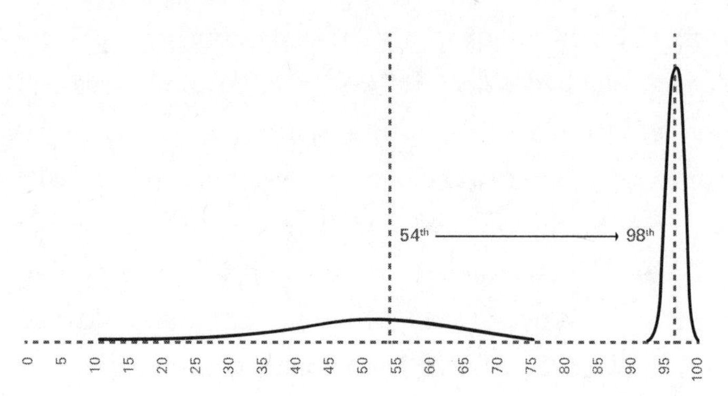

	min	max	mean	std dev
Sept. of Grade 5, 2008 (Pre-JUMP)	9%	75%	54%	16.6%
Sept. of Grade 6, 2009 (1 year of JUMP)	95%	99%	98%	1.2%

*Class percentile ranking based on results on the norm-referenced Test of Mathematical Abilities

who participated, fourteen received awards of distinction and the other three were close behind. Students who write the Pythagoras competition are generally in the top five percentile in achievement, but the average score for students in this (initially unremarkable) class was higher than the average for students writing the Pythagoras.

Now, this is just a case study. But Moreau's students were not an alien species, so their brains were likely similar to those of students in other regular classes. As well, JUMP has participated in larger studies and pilots that suggest that children are capable of much more than we expect. (For a summary of JUMP results, select "Programs" then "Research" on the jumpmath.org website.)

The most challenged ten-year-old student in Moreau's class improved her score on the TOMA from the 9th percentile to the 95th percentile after only one year. As the brains of ten-year-olds are less plastic than the brains of younger students, it seems reasonable to assume that Moreau's student could have achieved much more in grade five if she had been enrolled from an early age in a math program that fostered ability and promoted a growth mindset. Based on my observations of thousands of children, and the research I will discuss in this chapter, I believe the vast majority of grade five students (possibly as many as 99 percent) could, if they consistently had teachers like Mary Jane, do the work we expect of very strong students.

I visited Moreau's class to teach a lesson once. The students were all bursting with excitement about math and insisted that I give them harder and harder problems. At one point, they taught me something about patterns in division that I had completely forgotten. On another occasion, when I tried to meet with Moreau at the end of a lesson, her students demanded that she give them their bonus questions before she left the class—she had written them on the board and forgotten to remove the sheet of paper that was hiding them. I have seen this same collective engagement in many classes. I've seen students become so excited about math that they beg to stay in for recess to finish their work or ask for extra work for the summer. I even broke up a fight once by telling the instigator that if he didn't apologize I wouldn't give him his bonus question. And he apologized—for a bonus question in math!

Many people think that teachers will always be forced to make an impossible choice between helping weaker students keep up or allowing stronger students to move ahead. But Moreau's results show clearly that teachers don't have to make this choice. The lowest score on the TOMA after JUMP was at the 95th percentile, a full twenty points above the highest score prior to JUMP. It's hard to believe that teachers can help their strongest students fulfill their potential by helping the students who are initially weaker fulfill theirs at the same time. But Moreau's strongest students did better in grade five than they had ever done at school, in part because the whole class went further together. And in grade six, all of her students flew through the grade six math curriculum and then covered half of the grade seven curriculum. Some of her weakest students eventually outperformed the students who were initially stronger.

Moreau was able to shift the bell curve in her class so dramatically because she made all of her students feel like they could accomplish *roughly* the same things. In her classroom, students worked to compete against the problem, not each other. They got caught up in the excitement of their peers, and this excitement helped them to engage more deeply, remember what they learned and persevere in the face of challenges. They were encouraged to learn and love learning for its own sake, not because they were afraid of failing or wanted to be ranked higher than other students.

I believe that reducing the level of intellectual inequality in our schools (and workplaces) would do more to improve the world than almost any other social intervention we could invest in, not only because inequitable learning environments are extraordinarily unfair but also because they are inherently inefficient. Such environments are not good for *any* learners—including the ones at the top of the academic hierarchy—because they train learners to give up too easily or to exert themselves for the wrong reasons. They destroy our natural sense of curiosity and make our brains function in the most inefficient ways possible. And, as I will argue in chapter six, they also prevent us from developing productive mindsets. Fortunately, cutting-edge research in a number of fields suggests that math is a subject in which teachers can easily create more equitable and productive learning environments—even for older learners.

This Is Your Brain on Math

There is nothing Vancouver teacher Elisha Bonnis likes better than helping her grade five students see patterns in numbers or make connections between mathematical concepts. But for many years, she felt like a fraud whenever she had to teach math.

Bonnis struggled with math for most of her life. She first started to fall behind in the subject in grade three, after missing a few weeks of school because of bronchitis. There was no one at home who could help her catch up—she came from a family that was constantly on the move—and she was afraid to ask for help at school. As she fell further behind in math, she developed a chronic fear of the subject and her teachers began to treat her as if she was incapable of understanding the most basic numerical concepts. As she told a reporter for the *Vancouver Sun* during an interview about her experience with JUMP: "I thought it was only me, that I was the only one who didn't get it. Many times I was told I was just not a math person."[3]

Bonnis's struggles in math eventually began to affect her achievement in other subjects. She started skipping classes, failing tests and talking back to her teachers. After being expelled from two schools, she enrolled in an alternative school where she was able to finish her high school degree with high marks—except in math, as she had dropped the subject several years earlier. When she decided to become a teacher and applied to the University of British Columbia education program, she was horrified to learn that she would have to upgrade her math skills. As she recalls: "At UBC, I was again haunted by my deep and abiding hatred for and terror of math. I would study three hours each night, crying most of the time. I did pass, but once I became a teacher, in the first half of my career, I continued to feel totally inadequate when it came to teaching math, as if I were faking it. I just followed the textbook verbatim; I now know what a disservice I was doing my students."[4]

I met Elisha Bonnis when she attended a talk that I gave at the Vancouver school board in 2008. One of her colleagues, in whom she had confided about her fear of math, had persuaded her to attend the event. After the talk, Bonnis began experimenting with the online JUMP lessons; eventually she implemented the full program in her classroom. As Bonnis diligently worked her way through the lesson plans with her students, her anxieties about math began to fade away and she started to understand what she was teaching for the first time. Three years later, with a new-found confidence in her abilities, she enrolled in a master's-level program in math education at UBC. She completed her degree with high marks and now likes to mentor fellow teachers who are anxious about math.

I know many people who have discovered—sometimes rather late in life—that they have a talent for math and that they actually enjoy learning the subject. Among the hundreds of adults and teenagers I've worked with who thought they were "bad at math," I've only met a handful who weren't able to quickly catch up in whatever topic I taught them.

In chapter five, I will describe my work with Lisa, a teenager who was one of the most challenged students

I've ever met. In my first lesson with Lisa, I was surprised to see that she couldn't perform the most basic arithmetic or count to 10 by twos, even though she was in grade six. I soon found out from her principal that she was working at a grade one level at school. She had a "mild intellectual disability" (which means her IQ was supposedly around 80) and she had developed a paralyzing fear of math. After three years of weekly tutoring, Lisa told me she wanted to enrol in academic grade nine math. I was afraid she wouldn't pass the course, but she surprised me by skipping a year and finishing academic grade ten math in the same year.

People who think that math is an inherently hard subject sometimes draw an analogy between expertise in math and expertise in fields where people usually become competent performers only if they start learning when they are young, for example, being able to speak a language without an accent, consistently produce melodious notes on a violin, or perform a complex routine in gymnastics. According to this view, if a person doesn't display an ability with numbers fairly early in life, then they are probably not, as Elisha Bonnis was told, a math person. But research from a variety of fields – including cognitive science, neurology and even the foundations of mathematics – suggests that this analogy is flawed and that it is never too late for a person to learn math.

To give an example, if a person doesn't learn to speak a language before the age of six, then they are highly likely to speak the language with an accent, no matter how hard they try to sound like a native speaker. But recent research in childhood development suggests that there are no similar predictors of future success in math. In fact, for children, the strongest predictors of later achievement in math involve skills and concepts that every person will almost certainly develop, no matter how much they struggle in math in their early years or how delayed they are in acquiring these skills. These indicators involve very simple tasks that humans have evolved to perform with relatively little instruction, including the task of counting to 10, or of correctly associating a numerical symbol (1, 2, 3 and so on) with a quantity (for example, an array of dots) or a position on a number line and recognizing which of two numerals stands for a larger quantity.[5]

The research raises a puzzling question. If the skills that predict achievement in math are as simple as counting or matching a number with a quantity—things that almost everyone will eventually do—why should it matter whether a person acquires these skills at, for example, the age of four or the age of four and a half? Why should a child who learns to count six months later than their peers be at greater risk of enduring a lifetime of struggles in math? The research suggests that differences in ability in adults are not primarily the product of cognitive differences between individuals, because we all eventually learn the concepts that predict achievement.

I argue that the variation is mainly caused by a system of education that turns insignificant delays or struggles into life-changing differences. I will present evidence from psychology that suggests that our attitudes about our abilities (which we develop by comparing ourselves to our peers at school and at work) and our teachers' attitudes are more likely to prevent us from learning math as children or adults than any inherent difficulty in the subject.

The new research in child development is in line with a series of profoundly important discoveries that changed the course of mathematics over a hundred years ago. These discoveries eventually led to the development of digital computers and the communication technologies we rely on today. They also have significant implications for the way math should be taught. In the early 1900s, logicians proved that almost all of mathematics, including its more advanced branches such as calculus and abstract algebra, can be reduced to the same trivial concepts and procedures—such as the process of counting or grouping objects into sets—that predict success in math. Unfortunately, this news was never leaked to the general public, perhaps because mathematicians didn't want anyone to know that their subject, which is widely considered to be beyond the reach of ordinary brains, can be reduced to simple, logical steps that are accessible to any brain. In this book we will look at a number of examples of concepts—such as division with fractions—that baffle most adults but that are actually very easy to explain. I will argue that math is different from the other subjects we learn at school because, as logicians have shown, it is inherently simple.

People who believe that math is hard also tend to believe that structural features of the brain determine how much math a person can learn. Neurologists are only beginning to map differences between the brains of adults and teenagers who are good at math and those who aren't. They still can't say much about how the structure of the brain limits or enhances our abilities. But what they have found should give hope to anyone who wants to learn math as an adult.

People who are good at math tend to process math in a particular part of their brain (called the left angular gyrus) that helps them retrieve and use mathematical information more efficiently than people who are not good at math.[6] Interestingly, the information that math whizzes use to outperform ordinary people is—you guessed it—fundamentally simple. People who can activate their left angular gyrus are much better at retrieving basic facts (like addition and multiplication facts) and at reading the meaning of various mathematical representations (like graphs, diagrams and tables). Their advantage in math comes from not having to waste mental energy on basic processing tasks so that they can focus on understanding the underlying structure of problems. Studies have shown that people who are less competent in math can learn, through training, to activate the same areas of the brain that experts rely on when they do math.[7]

The neurological research is consistent with a parallel stream of research from cognitive science that shows that many abilities we consider innate can actually be developed through a method of learning called "deliberate practice." This research has shown that there are more and less efficient ways to learn math (or any subject) and that the methods of instruction that we currently use in our schools and our workplaces tend to be highly inefficient, as they cause "cognitive overload" in learners and fail to draw learners' attention, in an effective way, to the salient features of the material to be learned. For example, most teachers tend to introduce mathematical concepts with overly concrete or specific examples (often framed in stories that are designed to make the math seem "relevant"), but research suggests that these representations can actually prevent students from seeing the deeper mathematical structure of problems. In fact, mathematical concepts are often easier to learn when they are presented with less language and by means of more abstract representations.

Our ability to think abstractly, using mathematics, is one of our greatest human gifts. It may also be the gift that makes us most alike and that we share universally with other people. Mathematics has given us the power to create a vast array of technologies and to discern laws and patterns that govern the natural world. It has also allowed us to see behind the countless distracting and superficial features of reality that make the plurality of things seem different when, viewed from a more abstract perspective, they are often essentially the same. If every person's gift for thinking abstractly could be developed, we might also see that we have much more in common with our fellow human beings than we think. We could create a more equitable and productive society and improve our lives in ways that people who have never experienced the beauty or the power of mathematics can scarcely imagine.

Everyone should have a right to fulfill their intellectual potential, just as they should have a right to develop healthy bodies. We don't have to wait until we have recruited an army of superhuman teachers or invented some miraculous new technology to guarantee that right. In the last decade, cognitive scientists and educational psychologists have begun to uncover the mechanisms by which our brains learn best. They have gathered evidence that the significant majority of people can excel at and love learning when they are taught by methods I describe in this book. One of the most important questions of our time is whether we will act on that evidence.

CHAPTER 2: THE UNREASONABLE EFFECTIVENESS OF MATHEMATICS

Imagine how you would react if, on a routine visit to your doctor, you learned that you almost certainly have cancer (with a 90 percent probability). I've never received a diagnosis of this sort, but if I did I know my life would change in an instant. All of my immediate concerns and worries would undoubtedly seem trivial compared with the thought that, unless the disease was treated immediately and effectively, I might die prematurely and never see my family or friends again.

Now imagine how you would react if, several days after receiving your diagnosis, you learned that your doctor had made a mistake in interpreting your test and that there was actually only a 10 percent chance that you have cancer. In this situation, I'm fairly certain I would feel as if I had just received a pardon from a death sentence. I might resolve to change my eating or other lifestyle habits to reduce my risk of actually developing cancer—but otherwise, I expect I would carry on with my life in much the same way as I did before the first diagnosis.

I invented this story about the consequences of a medical error to illustrate the powerful impact that numbers can have on our lives. But the scenario is not completely fanciful. Doctors do misinterpret the results of cancer tests—more often than you might suspect—not because the tests are unreliable or ambiguous, but because they don't know how to calculate basic probabilities.

You need two pieces of information to estimate the probability that a patient with a positive result on a test has cancer: the level of accuracy of the test, and the percentage of people in the general population who have that type of cancer. For a given test, you would expect that every doctor would give roughly the same estimate, especially since a higher or lower probability could determine very different courses of treatment for the patient. But psychologist Gerd Gigerenzer, at the Center for Adaptive Behavior and Cognition at the Max Planck Institute in Berlin, has found that many doctors can't correctly determine a patient's likelihood of having cancer from a particular test.[1] Gigerenzer asked radiologists who had been performing mammograms for twenty to thirty years (including heads of departments) what a woman's chances of having breast cancer were if she received a positive result on a test that was 90 percent accurate. Rather shockingly, their estimates ranged from 1 percent to 90 percent. The real probability was roughly 10 percent.

Why would doctors sometimes overestimate the risk associated with a positive result on a test for cancer? Imagine that you are playing a game where players take turns spinning the pointer on a spinner like the one shown below, and you are hoping that, on your turn, the pointer will land on one of the grey regions. If you wanted to compute the probability that this will actually happen, you would have to count all of the ways that you could spin grey and then compare this number with the total number of outcomes on the spinner. Since there are 3 regions that are shaded grey and 9 regions altogether, the probability of spinning grey is 3 out of 9 or $\frac{1}{3}$.

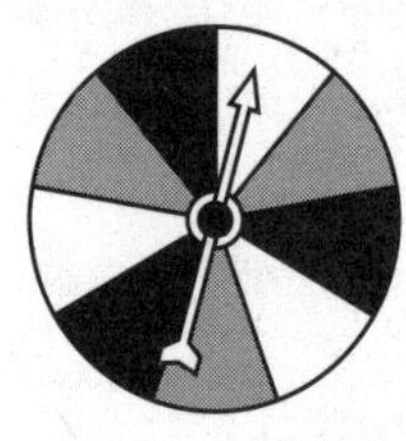

© Linh Lam

Now suppose that you somehow neglect to count all of the white regions when you are calculating the probability of spinning grey. In this case, you would count only 6 regions on the spinner (3 black regions and 3 grey regions), so you would conclude that the probability of spinning grey is 3 out of 6 or $\frac{1}{2}$ (which is a higher probability than $\frac{1}{3}$). Although it is highly unlikely a person would make this mistake with a spinner, this is similar to how doctors overestimate the risk associated with a test for cancer: they neglect to count some of the possible outcomes.

Suppose you have a test for cancer that is 90 percent accurate, and that out of 1,000 women only 10 typically have breast cancer. If you happen to test the 10, then on average 9 will test positive (because the test is 90 percent accurate). But that doesn't mean that your probability of having cancer when you test positive is 90 percent. We haven't counted all of the possible outcomes. We also need to consider how many people among the 990 who don't have cancer will test positive. Since the test is 90 percent accurate, it will be wrong 10 percent of the time. So, about 10 percent of the 990 people who don't have cancer (or 99 people) will get a false positive result. This means that out of every 1,000 people who take the test, about $9 + 99 = 108$ will test positive for cancer. But only 9 of those will actually have cancer. So the probability that you have breast cancer if you receive a positive result is only about $\frac{9}{108}$, which is 8.3 percent, or close to 10 percent.

Of course, 90 percent and 10 percent are just numbers. But when they represent two possible outcomes of a test for cancer, it's easy to understand the practical implications of a numerical error. Doctors who tell their patients that they have a 90 percent chance of having cancer, when the real likelihood is closer to 10 percent, can cause a high level of unnecessary stress in their patients, motivating them to seek treatments that they are unlikely to need and that may have detrimental side effects.

Because numbers are intangible and perform their intricate functions on scales we usually can't perceive—from the deadly codes that are inscribed in the DNA of viruses to the massive, element-making factories of the stars—we scarcely notice the impact that they have on our daily lives. But numbers are part of the fabric of our existence. They play a role in almost every decision we make—from the amount of debt we accumulate (individually and nationally) to the kind of treatment we choose to fight off a virus. There are many reasons why we would be wise to ensure that every member of our society who can vote, hold a job, sign prescriptions, sit on a jury, purchase commodities, build bridges, negotiate contracts, buy mortgages, bet on stocks, sell houses, consume energy or raise children has a basic working knowledge of numbers and mathematics in general.

Mathematics and Society

When we compare our present outcomes in education with those that cognitive science suggests are possible, or that teachers like Mary Jane Moreau can produce, it seems clear that even in the wealthiest societies we live in an age of intellectual poverty.

It's easy to see that our economy is less productive than it could be because so many people think they can't learn math. Corporate leaders often complain that they have trouble recruiting staff for skilled jobs or improving the productivity of their companies because they can't find people who know or are willing to learn the math required for technical or scientific jobs. A recent report by the President's Council of Advisors on Science and Technology estimates that there will be one million fewer STEM (science, technology, engineering and math) graduates over the next decade than US industries will need.[2]

People who struggle in math have trouble making sound decisions about their finances or about the economic policies they are asked to vote on. I've never heard anyone announce in public that they can't read a menu because they are illiterate, but I often hear people tell their friends (with some pride) that they can't add up a restaurant bill or figure out the tax. This lack of facility with basic math can have profound consequences. Ten years ago, the world was shaken by a recession that might have been avoided if people had understood what would happen to their monthly expenses if mortgage rates were to increase by a fraction of a percent. An increase of 0.5 percentage points doesn't sound very large, but when you are paying 2 percent interest, it represent a whopping 25 percent increase in your rate of interest (and roughly the same increase in your monthly payments)—as banking representatives pushing housing sales should have made clear. Our trouble with numbers may help explain why approximately ten million Americans and one million Canadians declared bankruptcy during the last decade.[3]

Many studies have shown a correlation between the quality of a person's education and the quality of their life. As it turns out, math can have an outsized effect on a person's life compared with achievement in other areas.

In 2007, Greg Duncan and a group of cognitive scientists analyzed the results of six large longitudinal studies of US students' academic outcomes from preschool to graduation.[4] They found that early math skills were significantly stronger predictors of later success at school than any other skills, including reading and attention skills. In 2010, two Canadian studies, one in the province of Quebec and one at a national level, replicated these findings.[5] These studies suggest that many of the positive impacts that a person's level of education can have on their quality of life are likely to be strongly dependent on their level of competency in math.

In 2005, sociologists Samantha Parsons and John Bynner used data from longitudinal studies of the British population to determine the effects of innumeracy in thirty-year-old men and women.[6] They found that people with "poor numeracy" were more than twice as likely to be unemployed as those with "competent numeracy." Men with poor numeracy, irrespective of their literacy levels, were also more at risk of depression, had little interest in politics, and were more likely to have been suspended from school or arrested by the police. Low levels of numeracy had even greater negative effects for women. Irrespective of their standard of literacy, women with low numeracy levels were less likely to have an interest in politics or vote, and were more likely to be in part-time jobs, in semi-skilled or unskilled jobs, or in a non-working household. They were also more likely to report poor physical health, to have low self-esteem and to feel they lacked control over their lives.

An earlier study by Bynner and Parsons also showed that people with poor numeracy tended to leave full-time education "at the earliest opportunity and usually without qualifications, followed by patchy employment with periods of casual work and unemployment."[7] Most of their jobs were low-skilled and poorly paid and offered few chances of training or promotion.

Mathematical competence is essential for making informed decisions in many areas of life. This is particularly true in decisions involving our health.[8] People who are less competent in math are less likely to understand the risks and benefits of screening or to take their medications with the correct dosage or frequency, so they have less successful outcomes in treatment than people who are more competent in math. Math can even have a surprising impact on a person's mental health. According to one study, depressed people who learn to activate their prefrontal cortex by doing mental math find it easier to control their thoughts about emotionally difficult situations.[9]

It's not hard to see that knowing math can help us make more informed decisions—but it can also help us solve problems and avoid creating them in the first place, by giving us the tools we need to think rationally and systematically about the risks and benefits of various political, environmental and economic policies. In a time when fake news and extreme opinions can spread like a contagion through social media, the capacity of the average citizen to engage in mathematical thinking is more important than ever.

At a press conference on January 11, 2016, a prominent US politician claimed that 96 million Americans were looking for work but couldn't find it.[10] When I read this claim, I thought it couldn't possibly be true, so I did the following simple mental calculation. I know the population of the US is about 300 million. I assume that about $\frac{2}{3}$ of the people in the US (or about 200 million people) are of working age. The figure 96 million is close to 100 million, which is half of 200 million. So if the politician's number was correct, then in 2016, about half of working-age Americans wanted jobs but couldn't find them. In other words, the unemployment rate in the US in 2016 was close to 50 percent! Sadly, very few people seemed to notice (or care) that this politician had made such an improbable claim, and only a few media outlets questioned the math behind it.

In the 1990s, politicians in New Jersey passed a bill that prohibited mothers on welfare from claiming tax benefits for children born after the bill was introduced. Two months later, when statistics showed that birth rates in New Jersey had dropped, some politicians claimed that their bill had caused this to happen.[11] They appear to have forgotten that pregnancies last for nine months, so the bill couldn't have had an impact in two months. (A longer-term problem with evaluating the effect of the bill was that some women on welfare stopped reporting births, because there was no advantage to doing so.)

If politicians were trained to think logically, they would know that you can't construct a valid argument unless you consider—dispassionately—all of the possible counter-examples to the claim you are trying to prove. And if they knew how to calculate probabilities and do basic statistics, they would be careful to weigh all of the factors that might be responsible for an event before claiming they know the cause. Our political debates would be much more reasoned and fruitful if people had to pass a course in basic mathematical reasoning before they could run for office.

When people try to use numbers to prove that human capacities like intelligence or talent in sports are determined by our genetic makeup, they often make elementary errors using ratios or percentages.

In *The Genius in All of Us*, David Shenk gives an interesting example of a claim that was circulated by media outlets around the world, even though it was based on an obvious mathematical error.[12] In the 2008 Olympics, Jamaican athletes shocked the US by winning 6 gold medals and 11 medals overall in track and field, compared with 25 medals overall for the US. This result was particularly impressive because the US population is approximately 100 times the population of Jamaica, so the US should have won many times more medals.

Sports commentators around the world immediately began to spread the story that Jamaican athletes had won a disproportionate number of medals because almost everyone in Jamaica carries a special variant of a gene (ACTN3) that regulates the production of a protein (alpha-actinin-3) that drives speedy and forceful muscle contractions.

In the US, ACTN3 is found in only 80 percent of the population, but in Jamaica it's found in 98 percent of the population. Ninety-eight percent sounds like a much bigger number than 80 percent, but to find out how many people actually carry the gene in a particular country, you need to multiply the percentage of people carrying the gene by the country's population. Given that the population of the US is about 100 times the population of Jamaica, I can be certain, without even performing a calculation, that multiplying the US population by 80 percent (or 0.80) will give me a much bigger number than multiplying the tiny Jamaican population by 98 percent (or 0.98). In fact, if you perform the calculation, you will find there are close to 100 times as many ACTN3 carriers in the US, because the US population is more than 100 times that of Jamaica's. So if the ACTN3 gene variant plays a significant role in producing Olympic medallists in track and field, the US should have won about 100 times as many medals as Jamaica did. (It appears that a national training program, started by an elite Jamaican sprinter, contributed to the success of the Jamaican athletes.)

Many people—including doctors and politicians who need to be numerate to do their jobs—have a sense of learned helplessness about numbers. Because so many of us struggled with math at school, we are easily fooled by arguments that involve erroneous math. We are also generally unwilling or unable to do simple calculations or to use elementary logic to analyze claims that involve numbers. Fortunately it's relatively easy to acquire the mathematical competencies we need to be more informed consumers of news and social media, navigate the complexities of our daily lives, or improve our prospects or performance in our jobs.

Darja Barr is a math professor who teaches a first-year math course to nursing students at the University of Manitoba. She has seen first-hand the toll that innumeracy can take on students who aren't prepared to take her class. Every year in North America, thousands of college and university students are forced to leave school or change their career path because they fail or receive low grades in entry-level math courses. Students who can't afford to drop out or opt for lower-paying careers take on heavy debt loads they can't repay because of their poor understanding of math.

Every year approximately twenty Indigenous students enrol in Barr's course, but a high proportion of these students fail or do badly because they don't have the mathematical background they need to succeed. In 2016, the average mark of the Indigenous students who completed her course was D plus. It broke Barr's heart to see so many students who were passionate about becoming nurses have their dreams shattered in her course. In 2017, in her spare time, she created and ran a one-week summer "boot camp" to help incoming students prepare for her course. She used JUMP lessons on number sense, ratios, algebra and fractions (available on the jumpmath.org website) as the basis for her curriculum. That year, the average mark of the Indigenous students who completed her course was B plus.

When you consider the impact that innumeracy can have on the life of a typical nursing student—including lost opportunities and dreams deferred for those who drop out, and mistakes on the job and barriers to promotion for those who barely pass—a week of extra work is a very small investment to make for the sake of a better future. I expect that most nursing students would welcome an opportunity to upgrade their math skills if courses like the one Darja Barr created were available.

And students in other disciplines would also benefit from this type of course.

In this book, I describe a set of basic mathematical competencies that the average person could learn in a relatively short "boot camp" of one or two weeks and that should be part of the mental tool kit of every adult. These competencies include the ability to do basic calculations with whole numbers and fractions; to compute and understand probabilities, ratios and percentages; to do simple algebra; to understand the meaning of elementary statistical terms; and to make estimates. I consider these basic mathematical competencies as a minimum requirement for productive citizenship. If they were expected of everyone who can vote, spend money or hold a job, it seems likely that our society would be more civil, productive and equitable. But there are deeper benefits of numeracy that people rarely consider when they weigh the importance of learning mathematics. If we truly understood the remarkable power of mathematical thought, and if we were able to experience, first-hand, the many ways that a mathematical perspective can enhance our imaginations and enrich our lives, we would make a much greater effort to develop the full mathematical potential of every person.

Mathematics and the Mind

In 300 BCE, the Greek mathematician Euclid formulated five postulates from which all of the geometric truths that were known at the time could be derived. As is the case with many important mathematical ideas, the postulates are so simple that they can be understood by a child. But they are also so powerful that mathematicians, scientists and engineers today are still finding new applications of the myriad truths that they imply.

Here are the postulates, in a modern formulation:

(1) A straight line can be drawn joining any two points.
(2) A straight line can be extended indefinitely.
(3) A circle with any radius can be drawn around a point.
(4) All right angles are equal.
(5) Given a straight line *A* and a point *B* that is not on the line, there is only one straight line that can be drawn through point *B* that doesn't intersect line *A* (or that is "parallel" to line *A*).

For many centuries, mathematicians were troubled by the fifth postulate, because it seemed more complicated and less intuitive than the other postulates. Between Euclid's time and the nineteenth century, many amateur mathematicians (and some professional mathematicians) claimed that they could prove the fifth postulate from the other four. But all of their proofs contained errors.

In the early 1800s, two innovative mathematicians decided to take a different approach to the fifth postulate.

Rather than trying to prove this axiom from the others, János Bolyai and Nikolai Lobachevsky independently set out to explore what would happen if they discarded the fifth postulate completely. To their surprise, they found that they could develop perfectly sensible and consistent geometries without the fifth postulate. Mathematicians soon realized that these geometries describe various ways we would experience the world if we lived on a curved surface or in a space of higher dimension that is curved. For example, if you lived in a universe with spherical curvature, and you travelled for a sufficient amount of time along a path that appears straight, you would eventually end up back where you started from. And if you lived on the surface of a large sphere (as you do), you might think (as some people still do) that you live on a flat surface. If the sphere was extremely large, you might not be able to tell that the surface is curved. But if you understood the mathematics of curved space, you could still deduce that if you *did* happen to live on a sphere, Euclid's fifth postulate would be false in your world.

On a flat surface, a straight line is a path that traverses the shortest distance between two points. On a curved surface, any shortest path between two points could be seen—from a more abstract perspective—as a kind of "straight line." These shortest paths will appear to be flat to a person who lives on the two-dimensional surface, but they will actually be curved in three-dimensional space.

On a sphere, there is only one shortest path between a given pair of points – and that path will always lie on a particular type of curve that mathematicians call a "great circle." One way to visualize a great circle is to imagine cutting a tennis ball into two equal halves along a plane that passes through the centre of the ball. The circular edge that you see when you look at either half of the ball is a great circle.

The Earth's equator is a great circle, as it divides the globe into two identical hemispheres. For any given pair of points on the Earth's surface there is only one great circle that joins those points. If you want to fly from the centre of Toronto to the centre of Sydney along the shortest route, you have to traverse the great circle that joins those two points. Airline routes often look curved when they are projected onto a flat map because pilots are following great circles to save time and fuel.

Because there are an infinite number of ways you can cut a sphere in half (along a plane through the centre of the sphere), there are an infinite number of great circles on the surface of a sphere. And each of these great circles will intersect every other great circle. That's why Euclid's fifth postulate is false on a sphere. Every shortest path or "straight line" (if extended) intersects every other straight line, so there is no such thing as a pair of parallel lines.

By way of contrast, on the saddle-shaped surface below, "straight lines" are parabolas and Euclid's fifth postulate fails for another reason.

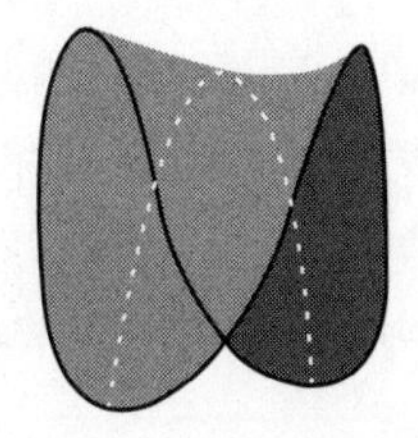

Given any line A and any point B that is not on the line, there is more than one straight line that passes through B that doesn't intersect line A. In fact, on a saddle-shaped surface, there are infinitely many lines that pass through B that are parallel to A.

One of the reasons mathematics is so effective is that it is abstract. Virtually all major progress in mathematics in the past two hundred years has been achieved because mathematicians have learned to see various mathematical entities, such as numbers, shapes and relations, more and more abstractly. A straight line on a flat plane and an equator on a sphere don't appear to have much in common, apart from the fact that they are both lines. But looked at from a more abstract perspective, they are each the shortest path on their respective surface and would be perceived as straight lines by someone living on either surface.

By investigating the geometry of curved space, mathematicians eventually developed the math that Einstein needed to posit the bizarre idea that matter can make space and time curve. And when astronomers showed, in 1918, that stars appeared to shift their positions during a solar eclipse, they not only proved that gravity could cause light to bend around the Sun, they also showed that the answer to an arcane mathematical question (about whether Euclid's fifth postulate was necessary) could underpin the most revolutionary theory that scientists have ever conceived of.

The kind of serendipity that led to the theory of relativity occurs over and over again in mathematics. The story often goes like this. A mathematician decides to investigate a problem that doesn't appear to have many (or any) practical applications, simply because they want to make a theory more elegant or beautiful or because they are curious. Many years later, their discoveries turn out to be exactly what a biologist, chemist, physicist or computer scientist needs to make a major conceptual breakthrough. Almost every modern branch of science, from genetics to quantum mechanics, is founded on concepts that were discovered by mathematicians fifty to five hundred years before anyone even conceived of the field. The physicist Eugene Wigner referred to this tendency for mathematics to consistently anticipate world-changing scientific and technological revolutions as the "unreasonable effectiveness of mathematics."

I can think of few living individuals who have demonstrated the practical power of mathematical thinking as clearly as the engineer and entrepreneur Elon Musk. Through his companies Tesla, SpaceX and The Boring Company, Musk has created a variety of revolutionary products and technologies. And he has inspired or forced many companies (particularly automotive and energy companies) to move more quickly to adopt environmentally friendly technologies and business models than they otherwise would have. Musk has attributed much of his success to his willingness to use mathematics to analyze problems from "first principles."

The Boring Company was founded recently, so it isn't as well known as Musk's other companies or his more famous inventions, like the hyperloop (which shoots commuter pods through tubes that have almost no air in them and therefore very little air resistance). The idea for the company was born when Musk, stuck in LA traffic, used mathematical first principles to analyze the problem of traffic congestion.

Traffic congestion costs our economy billions of dollars a year and can make commuters' lives extraordinarily unpleasant for several hours a day. But cities will rarely build underground tunnels to relieve congestion because the cost of drilling the tunnels has been prohibitive. That cost is determined by the volume of the dirt and bedrock that has to be removed. The volume can be calculated by multiplying the length of the tunnel by the cross-sectional area of the tunnel. This area, in turn, depends on the radius of the tunnel (that is, half the diameter). Musk realized that he couldn't change the length of a typical tunnel, but he wondered what would happen if he changed the radius.

Most of us learned the formula for the area of a circle (area equals the constant pi times the radius squared, or πr^2) in grade school. This formula implies that the cross-sectional area of a tunnel will increase exponentially as the radius increases, because the area is proportional to the square of the radius. If you have ever tried multiplying the numbers 1, 2 and 3 by themselves (or "squaring" the numbers), you know how quickly the square increases as you increase the size of the number. Musk conjectured that he could decrease the time required to drill a typical tunnel by a factor of 10 by reducing the radius of the tunnel. To compensate for the reduced width of the passage, he imagined shooting single cars through the tunnel at high speeds on sleds. Several days later, a company was born out of the elementary math that Musk did to pass the time while he was stuck in traffic. It's too early to tell how successful Musk's various companies will be—I, for one, wouldn't bet against him—but the mere existence of these companies, which were founded on mathematical intuitions, has already had a positive impact.

Mathematics can change the way our minds work by providing us with mental tools of incredible power. When we learn math, we learn to see patterns, to think logically and systematically, to draw analogies and to see beyond superficial differences using abstraction. We also learn to make inferences and deductions, look for hidden presuppositions, prove things from first principles, arrive at solutions by eliminating possibilities, develop and employ strategies to solve problems, and make estimates and "ballpark" calculations. And we learn to understand risk and cause and effect, and to have a sense of when data is significant or meaningless.

Not knowing how to think mathematically makes us, on average, less healthy, less financially secure, less innovative, less productive, less curious, less intelligent and less happy. It also makes us more prone to errors, more irrational, more superstitious and more susceptible to demagogues. Innumeracy damages our economy and degrades our environment. And there are other losses that stem from our failure to educate everyone according to their potential, but they are harder to put a price on.

Mathematics and the Soul

In 2004, I was invited to teach a demonstration lesson at a school in a very poor neighbourhood in inner-city London. While I was watching the children play at recess, I saw one fight after another break out across the playground. The children who weren't fighting formed rings around the ones who were to egg them on. Several children had to be carried off the playground because of their injuries, and the playground monitors couldn't stop the fights.

I had been invited to teach math to a behavioural class at the school. I told the class of eleven- and twelve-year-olds, who had been identified as difficult and even violent students, that when I was their age I thought I was dumb and wasn't very good at math. I said that if they didn't understand something, they could stop me and ask me to explain it again. It would be my fault, not theirs, if they didn't get something. Then I taught them to read binary codes, the strings of zeroes and ones that represent numbers for computers. The students seemed to think they were little code breakers and demanded longer and longer codes. When I performed a mind reading trick, they figured out the connection between the trick and the codes and wanted to come to the front of the class to do the trick themselves. On my third day with the class, when the teacher and I entered the room, the children cheered.

Not many people would associate math with social justice. But math is an ideal tool for changing how disadvantaged children think of themselves. I couldn't teach all of the students in a grade six class to read a story or essay with the same level of fluency and understanding in one lesson. But I *can* get an entire class to work on roughly the same level of math in a single lesson, because math can be broken into such trivial steps (or conceptual threads) and because it's possible to generate great excitement by making each step a little more challenging.

The confidence that disadvantaged students gain from succeeding in math can bolster them in other areas of their lives. When children see that they are capable of learning math, they begin to think that they can learn anything, because math is supposed to be hard. In London, in a pilot in which JUMP dramatically increased the pass rates for students at risk, one teacher reported that students with behaviour problems would reprimand others who misbehaved in math class because they were so engaged in their lessons. Another teacher wrote that her students had become "ballsy, independent problem solvers." In a pilot in Bulgaria, observers reported that students were happier, more engaged and more co-operative in classes that used JUMP. Supply teachers could tell if a class was using these learning strategies even when math wasn't being taught because the students were more engaged and curious and better at working together.

I've taught the binary code lesson in many classrooms, but I had one of my most memorable teaching experiences in a grade five class in Vancouver. Partway through my lesson, when the students were working on some questions that I had written on the board, I noticed a shy-looking boy, who seemed small for his age, hunched over his desk and rapidly filling up a piece of paper with calculations. When I glanced at his page, I noticed that he was writing down the numbers from 1 to 15 and translating them into binary codes.

How to Write a Number in Binary Code

Because we have ten fingers, our number system is built around multiples of ten. To a human, the number 1101 represents one thousand, one hundred, no tens and a one:

Thousands	Hundreds	Tens	Ones
1	1	0	1

A wire in a computer essentially has only two states—it either has electricity in it or not—so a computer can effectively only work with two symbols. That's why computers use a number system that only has two digits (0 and 1) where the place values are multiples of two. To a computer, the binary code 1101 represents one eight, one four, no twos and a one:

Eights	Fours	Twos	Ones
1	1	0	1

To find the number that this code represents, add any multiples of two that appear overtop of a one in the chart. Since eight, four and one each have a one underneath them (and 8 + 4 + 1 = 13) the binary code 1101 represents the number 13. Similarly, the code 1010 represents one eight and one two, or 10.

It's relatively easy to change a number given in binary code into a regular number, as shown in the figure above, but translating a number into a binary code is a little more difficult. I had taught the students how to translate binary codes into numbers, but the boy had figured out how to go backwards from the numbers to the codes. As I was just about to teach this to the students, I held up his paper and showed them his work. Then I challenged the class to figure out how to "crack the code" as he had done.

When the lesson was over, the teacher told me that she had been very nervous about letting the boy participate and had only decided to do so at the last moment. The boy, tragically, had been born with a very weak heart. His condition was so severe that he was not expected to live for more than a few years. He had always struggled in math and could become extremely anxious and frustrated when he encountered difficulties. To protect his health, the teacher had to be careful not to create any undue stress for the boy, and she had been afraid that he might have an anxiety attack in my lesson.

The teacher sent me a number of emails to tell me about the boy's progress. She said he "grew ten feet tall" in my lesson and that, when he went home that afternoon, he created his own book of binary codes and filled the pages with his calculations. He eventually told his parents that he wanted them to hire a tutor to help him catch up, because he had decided he was going to be a mathematician. He made remarkable progress in his tutorials and at school that year, and on his last birthday, he insisted on rushing home from his party because he didn't want to miss his lesson in math.

I remember many other students in that class as well, because they had all been bubbling with excitement during the lesson. One girl figured out that she could make a "computer" by putting some light bulbs in a row (with each bulb representing a different binary digit) and turning on the right sequence of bulbs to represent a given number. Several students discovered some interesting patterns (that I hadn't noticed) in the charts that I use for the mind reading trick. And one boy wrote me a rather unusual letter. When I read the date on the letter, I wondered for a moment if the boy had not been taught to write a date properly. Then I realized that it was in binary code. Here is the letter:

Wednesday, January 11010th, 11111011000

Dear John,

Thank you for coming to our class on January 11010th to teach us about binary codes. I enjoyed learning about how to read binary numerals. I always thought that translating big numbers like 2008 would be tough, but you made it so easy that my grade one brother, Will, was nearly a computer after I taught it to him your way!

He's already doubled up to fifty billion or something!

He was never very good at math, but he's memorized up to the eleven times tables in the last week. Are there any Jump Math books for grade threes? I think he is nearly at this level.

Help here would be appreciated.

From, Jasper

On the back of his letter, Jasper had written the powers of two (1, 2, 4, 8, 16 . . .), which are the basis of binary codes, in a column that went all the way down the page. At the bottom of the column, beside the number 4,394,967,246, he had written "Will's last double," and beside the number 33,554,432 he had written "Interesting number." Apparently Will or his brother had also found the powers of three up to 43,046,721, because they were also written on the page. I haven't given this much thought, but I am still not sure why 33,554,432 is an interesting number. But children are often able to see things that elude me.

Children who believe in their abilities can enjoy doing math as much as they enjoy making art or playing sports. They love to overcome challenges and are thrilled to discover or understand something that is beautiful, useful or new. They will happily spend hours solving puzzles, seeing patterns and making connections. But the majority of children will lose this sense of wonder and curiosity before they grow up, simply because, as a society, we expect so little of ourselves.

If children couldn't see any beauty in the visible world—in a snow-capped mountain or a sky full of stars—we would be concerned about the way they'd been educated or brought up. But there is a beauty in the invisible world of natural laws, in the elegant patterns and symmetries of every cell and every star, that can only be understood and appreciated through mathematics. Shouldn't we also be concerned when people fail to develop their capacity to see this aspect of the world?

There is no scarcity in the world of ideas. When someone understands an idea, its beauty is not consumed or used up. And when one person learns more, it doesn't mean another must learn less. But everything in our present system of education seems designed to make real knowledge scarce, to keep the deepest ideas out of the hands of all but a few. If they are lucky, students graduating from high school will likely believe that they have one or two talents and that the majority of subjects offered at school are either uninteresting or beyond their grasp. Twelve years is a relatively short time to close so many doors forever.

Rachel Carson, the famous environmentalist and author of *Silent Spring*, wrote: "Those who contemplate the beauty of the earth will find reserves of strength that will endure as long as life lasts."[13] And John Muir, who advocated for the establishment of the first US national parks, argued that nature can provide us with "infinite resources for happiness." If people were taught to see the beauty of the world on every level, and could appreciate its marvellous design with their intellects as well as their senses, then they might draw on even deeper resources of strength and happiness than the ones

Carson and Muir refer to. They would also understand how interconnected the world is and how quickly the small effects of individual choices can accumulate. Consequently, they would have a more refined sense of risk and be better equipped to protect the environment. They might even feel inspired to make more space for the inexhaustible and eternally satisfying beauty of the natural world in their lives—including the parts you can only see through math.

The hidden beauty of the world is accessible to anyone who can think. We don't have to compete or fight with each other to earn our share of this precious resource. If adults could see this beauty clearly, behind the fog of fear and confusion that people tend to feel when they learn math, they would want every child to see it too. They wouldn't let children lose their passion for learning or develop the flawed and destructive ways of thinking that are the products of intellectual poverty. To recover our capacity to see the world clearly, with our intellects as well as our senses, we need to start by examining the myths about ability and intelligence that have stopped so many people, including myself, from realizing their full potential in mathematics and the sciences.

CHAPTER 3: BECAUSE YOU GET THE RIGHT ANSWER

When I was growing up in the 1960s, my parents purchased a collection of books from Time-Life that sparked my interest in the sciences. The books, full of beautiful images and intriguing ideas, covered topics like the planets, the oceans and the animal kingdom. My favourite book was called *The Mind*. In the middle of the book I found a painting of seventeen geniuses who had all made profound contributions in their fields. Above each portrait, the artist had sketched, using a very ornate script, the subject's name and IQ.

I would go back to that painting frequently, throughout my childhood, and pore over the numbers the way other children might pore over baseball stats. I was fascinated by the idea that for every individual there was a number that would tell you everything you needed to know about their intellectual capacity. Because the IQ scores in the painting were featured in a book on science, I assumed that they were very precise and that Goethe was a much more important thinker than Galileo because his score was a full 25 points higher.

I read other books on the brain, including a book I found on my sister's bookshelf on giftedness in children (she was studying psychology at university). I didn't understand everything in the books, but two facts always came through very clearly: IQ was something you inherited from your parents, and it was absolutely immutable. That meant I would have to live with whatever intellectual capacity I was born with for the rest of my life; no amount of effort would change my IQ score by a single decimal point. As a child, I often fantasized about inventing or discovering something new, like the geniuses in *The Mind*. So the thought that my intelligence had been determined at birth and might be too low for me to do anything original or interesting was almost as disturbing as the Calvinist view that there is nothing you can do to secure a place in heaven: the chosen are preordained and the rest are condemned to eternal torment no matter how hard they work to save themselves.

Over the past few decades scientists have discovered that many qualities of mind and behaviours that are critical for success in school and in life are not measured by IQ tests: these include creativity, the ability to persevere and defer gratification, the ability to collaborate with others, the willingness to use trial and error to find solutions to problems, and the willingness to be guided by evidence and facts and to carefully follow basic principles of logic and rationality (rather than biases and wishful thinking) in engaging with the world. In chapter six, "The Psychology of Success," I will present various strategies that adults can use to develop more productive habits of mind and that teachers can use to help their students become more resilient and efficient math learners.

Scientists have also begun to generate evidence that "fluid intelligence" (the intelligence we use to solve novel problems) is malleable and can improve through training. For example, several years ago psychologists Allyson Mackey and Silvia Bunge asked a group of seven- to ten-year-olds to play commercial board games that rely heavily on reasoning (such as Rush Hour, where players need to figure out how to escape from a traffic jam while still obeying the rules of the road). After playing the games for an hour a day, two days a week, for eight straight weeks, the children increased their scores on a reasoning test by more than 30 percent and their IQ scores by 10 points on average.[1] In the 1990s, psychologist James Flynn discovered that IQ scores had increased steadily over the previous fifty years, rising roughly three points per decade on average in many countries, with the most significant gains occurring for fluid intelligence. Psychologists have offered many explanations for the "Flynn effect"—including higher levels of education in the general population and more cognitively demanding jobs—but whatever the cause, this phenomenon suggests that people can become more adept at the skills measured on IQ tests.

In 1997 when JUMP was still a tutoring club in my apartment, I had the opportunity to act in the movie

Good Will Hunting. The interior scenes of the movie were shot at the University of Toronto and the writers of the movie, Matt Damon and Ben Affleck, had asked the math department to recommend a consultant to look at the math in the script. Because of a miscommunication with one of the producers, I didn't end up working as the math consultant. My old physics professor, Patrick O'Donnell, got the job. But the director cast me in the role of a graduate student, Tom, who is jealous of the main character, Will Hunting. I loved working with the artistic team, who were all extraordinarily generous and encouraging. But after several days of shooting, I was feeling uneasy about being in a movie that plays heavily on the idea that geniuses like Will are born and not made. I asked the writers and the director if I could add a few lines to the movie that would provide a different view of ability. They understood my concern and were open-minded enough to let my character Tom say the following lines: "Most people never get a chance to see how brilliant they can be. They don't find teachers who believe in them. They get convinced they're stupid."

One way to get people to examine their beliefs about their intellectual abilities is to show them how easy math can be when it's taught well. Math is a particularly effective tool for changing mindsets because the majority of people still believe that math is an inherently difficult subject. They tend to talk about mathematical talent the way psychologists once talked about IQ. They assume, for instance, that success in math is an extremely strong indicator of a person's intellectual capacity and that a person is either born with mathematical ability, like Will Hunting, or not. I have encountered many parents who told me they didn't expect their children to do well in math because they had not inherited a talent for it themselves.

The Expert Mind

The view that children need to be born with ability, or develop ability in childhood, to excel in various intellectual pursuits, like chess or math or physics, is deeply prevalent. The most extreme form of this idea says that intellectual ability is hard-wired into the brain by our genes and can only be developed by people who have the right kind of genes. Fortunately this simplistic view of genetics began to lose its currency at the end of the last century when scientists discovered that genes are controlled by another system of molecules (that are attached to our DNA) and that this "epigenetic" system can cause genes to express themselves or remain dormant in ways that are heavily influenced by the conditions in which a person lives—or by their "environment." As David Shenk says in *The Genius in All of Us*:

> *Rather than finished blueprints, genes—all twenty-two thousand of them—are more like volume*

> *knobs and switches. Think of a giant control board inside every cell in your body.*
>
> *Many of these knobs and switches can be turned up/down/on/off at the same time—by another gene or a minuscule environmental impact. This flipping and turning takes place constantly. It begins the moment the child is conceived and doesn't stop until she takes her last breath. Rather than giving a hardwired instruction of how a trait must be expressed, this process of gene–environment interaction drives a unique developmental path for every unique individual.*[2]

Scientists have also discovered that the architecture of the brain is not—as they used to think—fully determined in childhood and that the adult brain is constantly creating new connections between individual neurons and networks of neurons in response to new experiences and learning. In a famous study published in 2000, Eleanor Maguire found that the hippocampus (the part of the brain that is responsible for processing spatial information) is more highly developed in London cab drivers, who must navigate a vast and notoriously complex network of streets to earn a living, than in bus drivers, who always travel the same routes.[3] Since then, studies in many fields, including music, sports and medicine, have shown that when a region of the brain is repeatedly activated through practice—as a person develops a new skill or acquires new knowledge—the brain's structure can change dramatically.[4]

In the early 1990s, psychologist Anders Ericsson began a series of studies that have shed new light on the ways people develop extraordinary abilities. In one study, he conducted extensive interviews with a group of advanced violin students at the Berlin University of the Arts to see if he could find anything in their musical histories that would explain their abilities. He gathered data on their teachers and on the number of lessons they had taken, the age at which they started playing, the number of competitions they had participated in, the number of hours they had spent playing solo as opposed to playing in groups, and the number of hours they had spent listening to music or studying theory. When he compared the data to the professors' ranking of the students, Ericsson was surprised to find that there was only one factor that separated the extraordinary violinists from the ones who were merely good: the most talented players had spent significantly more time practising their instrument. Since then many other studies have confirmed that the most talented people in any field invariably spend much more time practising than their peers.

Ericsson and other psychologists have also found that in some fields (such as chess, musical performance and competitive sports), expert teachers and practitioners have developed, over the course of centuries, methods of practice that are much more efficient than the methods that a non-expert would typically use to improve their performance in those fields. For example, to improve their performance in chess, people who don't know how to train efficiently will typically play the full game over and over. But psychologists have found that players advance much more quickly by taking a more incremental and focused approach: this might involve playing a mini-game, with just a few pieces, repeatedly until you can see the best move or moves, analyzing a single position in depth, studying the moves of master players and memorizing effective moves and strong positions. In chapter five, I will discuss a number of educational ideas that are very popular and that have led educators to develop and promote inefficient methods of practice that make math appear to be much harder to learn than it actually is.

Psychologist Carol Dweck has shown that people's "mindsets" play a significant role in academic achievement.[5] Some people have "fixed" mindsets; they believe that they will only excel in a subject if they are "smart" or have an innate talent or natural ability. Others have "growth" mindsets; they believe their success depends on their willingness to work hard and persevere. Dweck showed that students with growth mindsets do better at school because they have more productive habits of mind. Students with fixed mindsets tend to avoid work and will give up easily if they have to struggle to learn something. In their minds, if you have to work, it shows that you don't have natural talent. Dweck also showed that students do better when teachers praise them for being persistent and working hard rather than for being smart or gifted. After watching a video of a JUMP lesson, Dweck said: "JUMP Math already implicitly incorporates a lot of growth mindset principles . . . the kids are moving at an exciting pace, it feels like it should be hard but it's not too hard for them . . . they all have the feeling of progress and they all get the feeling that 'I can be good at this.'"

In the lesson that Dweck watched,[6] I gave the students a series of challenges that they could always overcome, because all of the skills and concepts they needed to succeed were embedded in the challenges. Rather than reinforcing the idea that only a small group of students with special talents can do math, the lesson showed every student what they could accomplish if they worked hard and didn't give up.

Twenty years ago, when I was making plans to start a free tutoring program in my apartment, I wasn't sure what subjects I wanted the tutors to teach. I had taught a variety of subjects myself—reading, creative writing, philosophy, critical thinking, math and science—and I could see how disadvantaged students might benefit from getting help in any of these areas. However, I eventually settled on math, precisely because it is considered to be such a difficult subject and because I knew that it is actually very easy to teach.

In the lesson that Dweck watched, all of the students in the grade six inner-city class where the lesson took place were able to solve the same problems about perimeter by the end of the lesson. But the lesson got off to a rocky start when I asked the students to copy a simple L shape (shown below) from the board and write the lengths of the sides on their pictures.

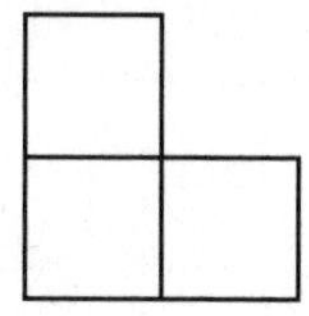

When I ask teachers to predict where students might neglect to label a side in this shape, many will venture that students are most likely to overlook one of the corner sides. And this is exactly what happened in the first few minutes of my lesson. When I checked the students' work, I was surprised to see that a fifth of them had written a single numeral 1 beside the corner sides (when they should have written two 1s because there are two corner sides).

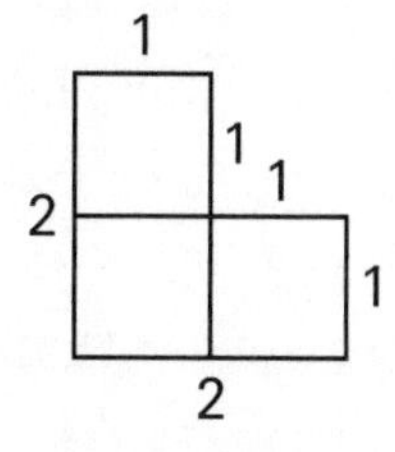

Correct labelling

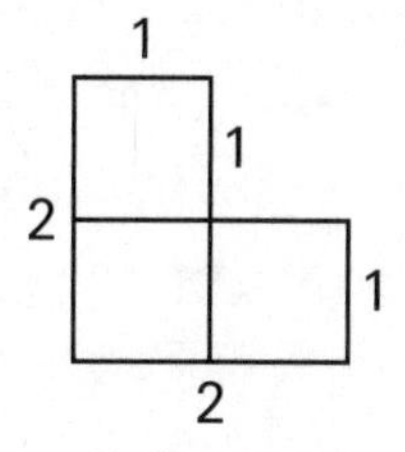

Incorrect labelling

I had to spend a few moments helping these students correct their error, while the others worked on some bonus questions. But soon all of the students were able to find the perimeters of more complex shapes. They were also able to solve some puzzles that involved adding sides to shapes so that the area of the shape increases but the perimeter of the shape stays the same or even decreases.

Later in the lesson, I asked students to draw as many rectangles of perimeter 12 (with whole-number sides) as they could. I was surprised to see that some of the students, after making the width of the rectangle 1 unit long, drew another side of length 11.

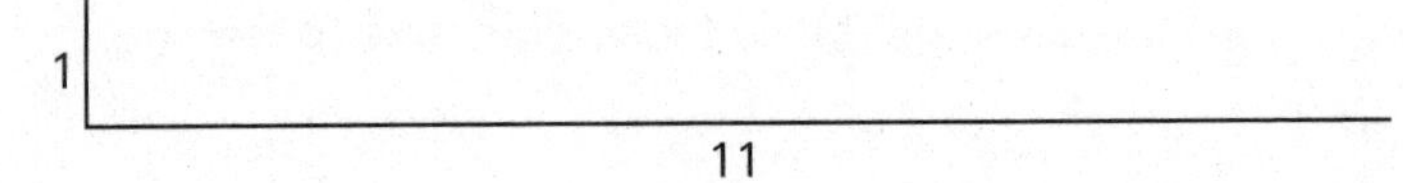

It was clear that these students had no idea how the perimeter wrapped around the shape—they didn't realize that if they made the longer side 11 units long, they would use up all of the perimeter. I had to give them time to draw various side lengths until they saw that, if the width was 1, the length could only be 5 units long (or the perimeter would be greater than 12). Soon I was able to give them more complex problems where they had to calculate the length of various missing sides given the perimeter of a shape. By the end of the lesson, all of the students were working on the same problems.

If I had been teaching a different subject, especially one that required strong reading skills or extensive background knowledge, I might have had trouble getting all of the students doing the same work, even if I had many lessons to work with them. But in math classes I can usually get all of the students working on the same material in one or two lessons. There are usually only a small number of skills or concepts that I need to review, or misconceptions that I need to catch, in order to include everyone in the lesson.

The methods of teaching that I discuss in this book work well in any subject, but, for reasons I will elucidate more fully in chapters four, five and six, math is the subject where these methods can have the most immediate and deepest impact. There is no other subject where we can mitigate differences between students so quickly or build productive mindsets so easily. Rather than thinking of math as a subject that is accessible only to the brilliant few, we need to recognize it as a powerful educational tool for creating a more equitable society.

Why You "Flip and Multiply"

In chapter one, I described a body of research from a variety of fields—including logic, cognitive science and early childhood development—that lends support to my belief that math is an inherently simple subject and that high-level mathematical thinking is founded on very basic cognitive functions. A recent groundbreaking study by French neurologists Marie Amalric and Stanislas Dehaene provides even more evidence to support this claim.[7]

Amalric and Dehaene asked a group of mathematicians and non-mathematicians to answer a series of questions about complex mathematical and non-mathematical topics while they used functional magnetic resonance imaging (fMRI) to track which parts of the subjects' brains were activated. When the subjects thought about the non-mathematical questions, the scans revealed that they all used areas of the brain that are normally involved in language processing and semantics. But when the mathematicians thought about the mathematical questions, the scans showed, rather surprisingly, that they activated the same neural networks that young children use when they think about math. Based on these results, the researchers speculate that the complex machinery of advanced mathematics is built out of the simple intuitions of number and space that we share with many primates. According to Amalric, the results show that "high-level mathematical reflection recycles brain regions associated with an evolutionarily ancient knowledge of number and space."[8]

To understand why so many people struggle in math, when evidence from so many fields suggests that mathematics should be accessible to everyone, we need to look closely at the systemic problems that make it so hard for teachers to nurture their students' full potential. Until recently, very few of the math programs and resources in use across North America have been tested in rigorous scientific studies, so educators and parents have often been seduced into adopting instructional approaches that sound very progressive and appealing, but that lack strong empirical evidence.

For example, when parents, teachers or administrators choose resources for students, they will try to find materials that they believe are engaging or interesting for them. But the choices they make are rarely informed by rigorous research. In 2013, psychologists Jennifer Kaminski and Vladimir Sloutsky of Ohio State University taught two groups of primary students to read bar graphs using two different types of graph: one type had pictures of stacked shoes or flowers, and the other, more abstract graph had solid bars (see page 76).

The researchers asked teachers to say which kind of graph they would use with their students. The significant majority chose the graphs with pictures because they were more engaging and represented the objects in the problem. However, the study showed that students learned better from the grey monochrome bars. Students who learned with the bars were better at reading graphs when the scale of the graph changed to reflect some multiple of the number of objects. Students taught with pictures tended to be distracted by counting the objects and so did not look at the scale on the graph.[9] In chapters four to six, I will give examples of

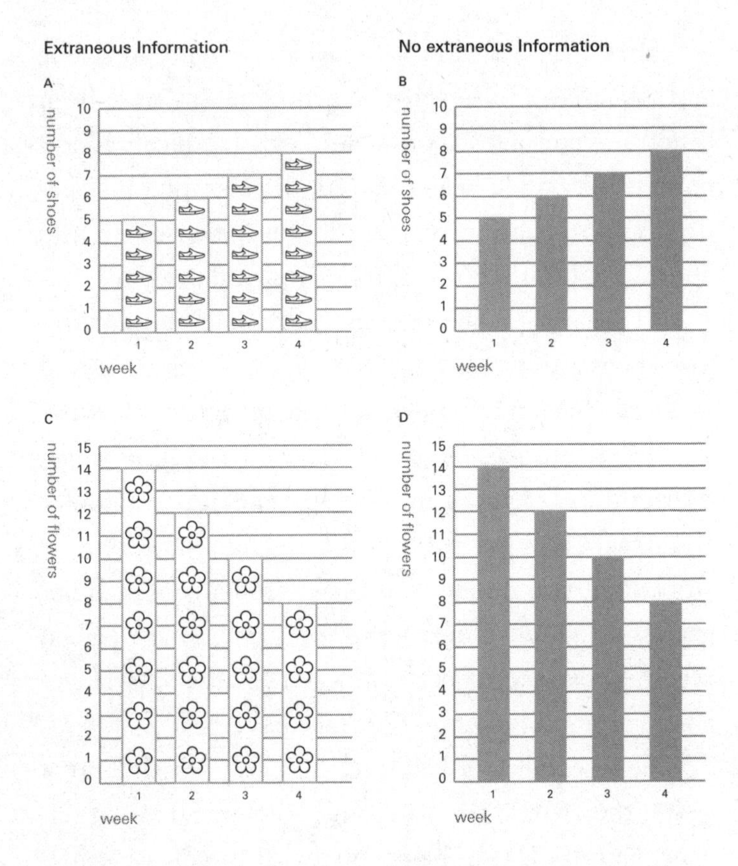

other popular approaches to teaching that actually inhibit learning (usually by overwhelming the brain with too much new information at once).

Teachers are often blamed for low test scores, ineffective lesson plans and failing schools, but in my opinion they are not ultimately responsible for these problems. In fact, I believe teachers should be commended for helping their students as much as they do, especially when so many of the resources and methods of instruction that they are required to use have been deemed ineffective by cognitive scientists. When teachers are exposed to research in cognitive science (which they rarely encounter in their professional training) and offered an opportunity to improve their practice, I have found that they will usually seize it.

Many teachers, particularly at the elementary level, will admit that they don't understand math deeply or love teaching the subject (especially when it comes to topics like fractions or algebra). In a keynote address at the Calgary City Teachers' Convention, I asked seven hundred teachers why, when you divide seven by two thirds, you are allowed to flip and multiply (or why seven divided by two thirds is the same as seven times three halves).

$$7 \div \frac{2}{3} = 7 \times \frac{3}{2}$$

Someone in the audience yelled out, "Because you get the right answer."

Over the past ten years I've asked hundreds of teachers why, when dividing by a fraction, you can invert the fraction and multiply. Only a few have been able to give me a simple explanation. The vast majority will admit that they learned this procedure as a rule that they never understood. If they *do* know an explanation, it is usually a very complicated one that involves multiplying both terms of the division statement by a fraction. Unfortunately, this explanation simply replaces one mystery with another, as most teachers don't know how to explain multiplication by a fraction either.

Because very few people understand why they "flip and multiply" when dividing by a fraction, I often start with this topic when I want to convince someone that math is not as hard as they think. I encourage you to follow along with me here and try to figure out for yourself why the rule works.

If you want to relearn math as an adult, it helps to go back to the level where you first started to struggle or feel confused. For many people, this would mean going back to grade three, because there is an ambiguity in every division statement that many teachers fail to notice or elucidate for students. If I ask you to divide six things by two ($6 \div 2$), what would you do with the 6 objects?

Some people will divide the 6 things this way:

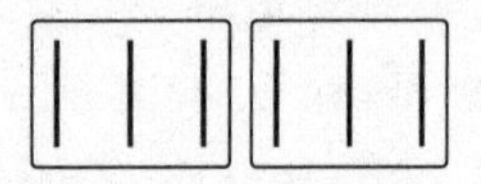

But, unfortunately for children in grade three, this is not the only possibility. Some people might divide the 6 things this way:

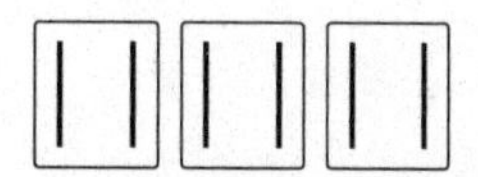

Both answers are correct: the division statement (6 ÷ 2) is ambiguous.

In the first case, what does the 2 refer to in the picture?

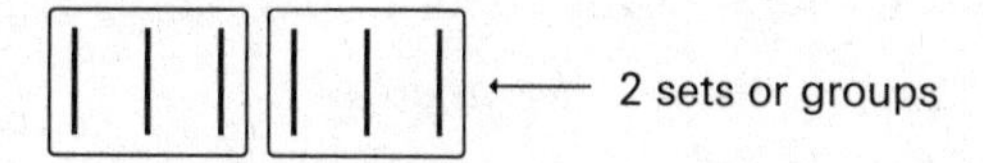

In the second case, what does the 2 refer to?

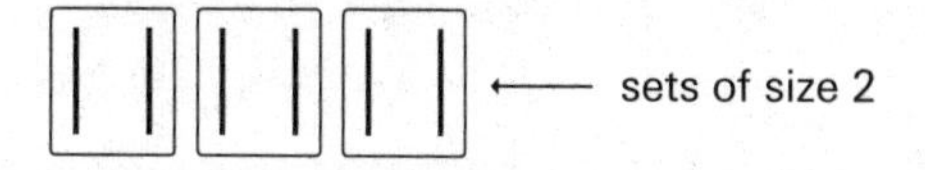

To know what the 2 in the division statement 6 ÷ 2 means, you need a context; you need to know whether you have been given the number of groups or the size of the groups. The same ambiguity holds no matter what model you use. For instance, suppose you have a chocolate bar with 6 pieces. What does 6 ÷ 2 mean? It can mean divide the 6 pieces of chocolate into 2 groups:

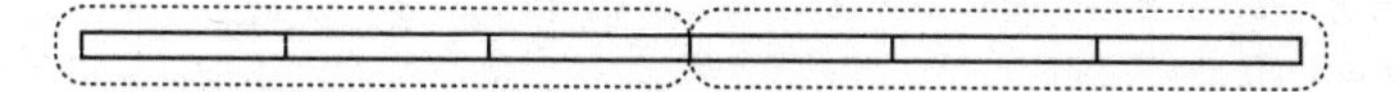

Or it can mean divide the 6 pieces into groups of size 2:

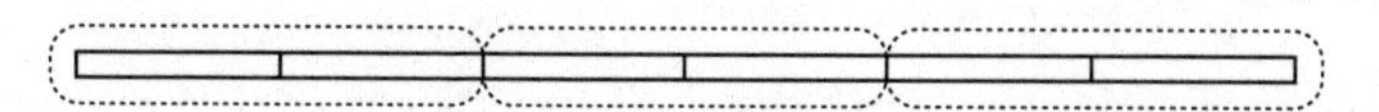

To reiterate: you can either interpret the divisor (in this case the 2) as the *number* of groups you want to divide 6 into or the *size* of the groups you want to divide 6 into.

Now consider this division statement:

$$6 \div \frac{1}{2}$$

I find it's easier for young students to understand this statement using the "size of the group" interpretation, rather than the "number of groups" interpretation. (It takes a little work to get students to understand what it means to divide six into *half* a group.)

Under the "size" interpretation, the statement $6 \div \frac{1}{2}$ means: divide 6 into groups of *size* $\frac{1}{2}$. To understand this, it helps to start with a model and also to identify what the unit is (you have to know what "one" looks like in the model). We can use a number line to represent a chocolate bar with 6 pieces.

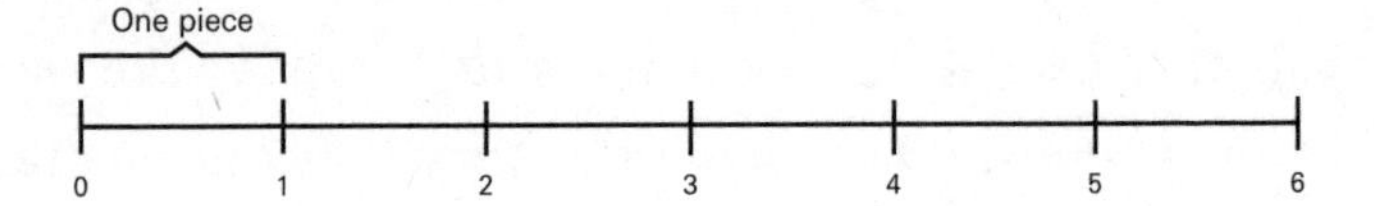

If I wanted to divide the first piece of chocolate into half-sized pieces, how many pieces would I get? Hopefully you can see that the answer is 2.

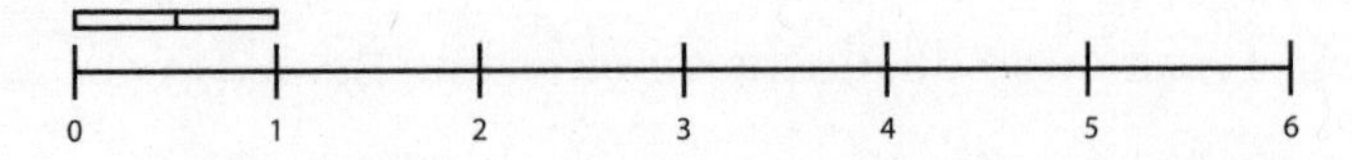

Now, if I divide up the rest of the pieces in the same way, how many half-sized pieces will I have altogether? As you can see, it's 12.

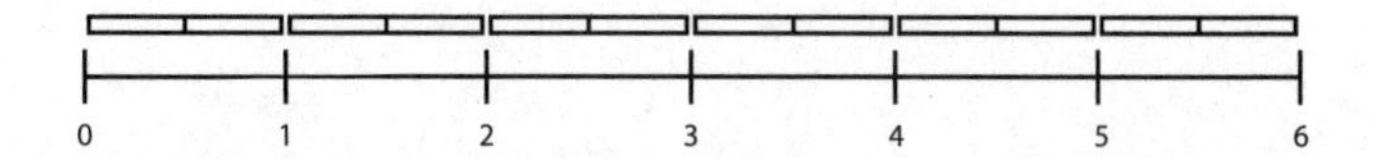

So six divided by a half is twelve.

$$6 \div \frac{1}{2} = 12$$

Now, suppose I divide six by a third ($6 \div \frac{1}{3}$). How many third-sized pieces can I divide the first piece into? (3)

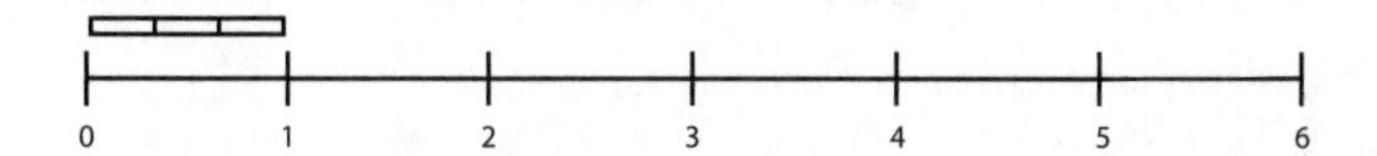

How many third-sized pieces will be in the whole chocolate bar? (18)

When I am teaching young students, I often say: "Now you're in big trouble. What if I gave you a question I couldn't even draw: $6 \div \frac{1}{100}$?" Students get very excited by these kinds of challenges, especially when they are in a group and can show off in front of their peers. The excitement helps focus their attention so that they can easily see the answer.

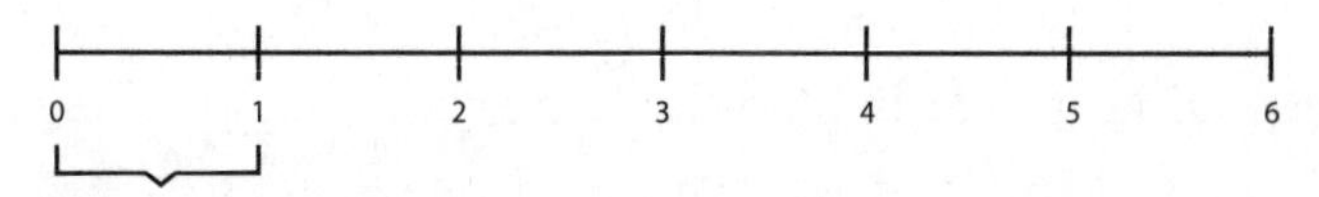

there are 100 pieces of size $\frac{1}{100}$ in one piece

so $6 \div \frac{1}{100} = 600$

Now I ask: Do you notice a pattern?

$$6 \div \frac{1}{2} = 12$$

$$6 \div \frac{1}{3} = 18$$

$$6 \div \frac{1}{100} = 600$$

In each case, what operation did you use to find the answer? You multiplied the dividend (6) by the denominator (2, 3, 100) of the fraction. (The denominator tells you how many pieces you want to subdivide each piece of chocolate into.)

Now I hope you can see why, when you want to divide a whole number by a fraction (with numerator equal to one), you flip the denominator and multiply. If you would like to know why you flip and multiply when the numerator of the fraction is not equal to one, you can watch the video "JUMP Math: Dividing by a fraction: how does flip and multiply work?" on YouTube.

When adults see how easy it is, through a series of Socratic questions and incremental challenges, to learn something that baffled them at school, they feel a little smarter. But they still may not be ready to believe that they could do higher-level math. They assume that there is a hidden depth to mathematics that can only be plumbed by a different kind of brain. In particular, they refuse to believe that they could ever develop a talent for problem solving.

Many adults struggle with the following elementary problem: "A person is standing 2,152nd in line and a second person is 1,238th in line. How many people are between them?"

Most people will subtract to find the answer, but if you ask them how they know their answer is correct, they often won't be able to say. I happen to know that this approach will give the wrong answer, but not because I was born with an innate ability to see this. As a mathematician, I know basic strategies for solving problems. I would never tackle a difficult or complex problem head-on if I could create an easier version of the problem and solve it instead.

In this case, I would draw or imagine five people in line and test whether subtracting the position numbers of two of those people actually gives the number of people in between. If a person is 4th in line and another is 2nd, then subtracting the positions gives the number 2, but there is clearly only 1 person in between. Subtracting always gives an answer that is one too high. Students who receive basic training in problem solving can soon experience the surge of pride that comes with solving complex and difficult-looking problems. In the next chapter, I will look more closely at the strategies that are most effective for solving mathematical problems and show how easy it is to adopt these strategies and to train people to use them.

Some areas of math require special skills (like the ability to visualize and mentally transform two- and three-dimensional shapes) that, until recently, appeared to be inborn. But in 2013, a meta-analysis of two decades of work on spatial training by D. H. Uttal revealed that spatial reasoning can be improved through a wide variety of activities across all age groups.[10] These activities include puzzle play, video games (like Tetris), block building, and tasks involving art and design. Other, more recent, studies have confirmed that children and adults can develop their abilities to mentally visualize and manipulate spatial images by playing readily available games.

The problem-solving strategies I describe here can be used in a wide variety of domains. When I was doing research for this book, I encountered the following problem: Imagine folding a sheet of paper three times, as shown in the top row of the picture on the next page, and cutting a corner off the paper. If you unfolded the paper, what would it look like (picture A, B, C or D)?

To solve this problem, I first tried the strategy of solving a simpler problem, which I described earlier. I imagined folding a sheet of paper once and attempted to visualize what various cuts would look like when the paper was unfolded. This was relatively easy, but unfortunately I was still two folds away from solving the problem. Then I remembered another strategy, which involves working backwards from an answer or end result. This strategy can be used to solve a wide range of problems.

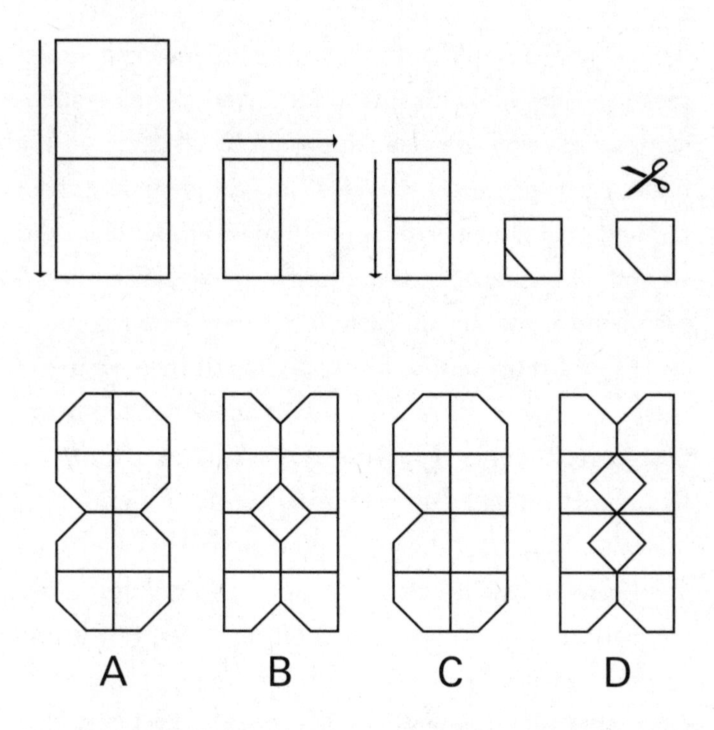

Rather than trying to make sense of the sequence of folds from the beginning, I looked at the final picture in the sequence (the one with the scissors over it). I found that I could unfold that shape in my imagination and keep track of the cuts working backwards.

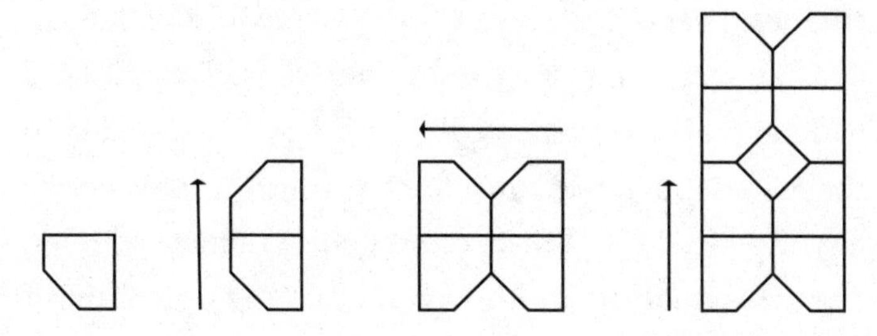

This is one example of the many techniques and strategies I rely on in my creative writing and my mathematical research that are also supported by research on the brain. By using these methods I can now learn new material and generate new ideas much more easily than I did when I was a teenager struggling at school, even though my memory and my ability to focus were better then.

When I returned to university to study math in my thirties, I had a significant advantage over the other first-year students. I had been working as a tutor for five years, so I had a deep knowledge of the high school material. I had also done a good deal of work getting a head start on the university material. In my first term, I consistently got high marks on my tests, but in my second term I failed a test in group theory (a branch of algebra) badly. I remember being virtually bedridden for a day when I received my mark. I thought that I had reached the limit of my abilities and that I would have to abandon my dreams of being a mathematician. That's when I decided to apply the insights I had gained from reading the letters of Sylvia Plath to my studies in mathematics. I set out to master the concepts in small, manageable steps and practised and reviewed whatever I had taught myself repeatedly, until I eventually learned the material that had caused me so much trouble on the test. By the time I finished my undergraduate courses, I knew enough group theory to teach the course that I had struggled in.

It's relatively easy to convince children that they have ability in math: in one or two lessons, I can usually build their confidence to the point where they are willing to make an effort to learn. But it's much harder to convince an adult to make that effort. Because of our experiences at school and an ingrained sense of the intellectual hierarchy, we all have fears and insecurities that cause us to give up too easily and to assume that every difficulty is an unbreakable ceiling for our talents. That is why it's so important for adults to learn about how their brains work, so they can develop productive habits of mind and distinguish between methods of instruction that make learning easier and methods that cause cognitive overload and make math and other subjects seem so hard.

Alfred Binet, who created the first intelligence test, once wrote: "Some assert that an individual's intelligence is a fixed quantity which cannot be increased. We must protest and react against this brutal pessimism."[11] One hundred years later, we are still suffering from the effects of this brutal pessimism, in ways that we are hardly aware of. It took me many years to learn that my intellectual and artistic potential is not determined by my IQ or the failing grades I sometimes received at school. But the majority of children, even in the most affluent schools, will not be lucky enough to develop a more optimistic view of their future as learners. To eradicate intellectual poverty, we must move beyond the idea that people have vastly different natural talents and inclinations in the arts and sciences. And we must stop reinforcing these ideas by using methods of teaching that overwhelm the brain and cause the majority of students to lose faith in their abilities at an early age.

In the next chapter, I will demonstrate a method of teaching and learning I call structured inquiry[12], which is a form of deliberate practice adapted to mathematics. I will show how even people who lack confidence in their abilities are able, through this method, to learn to solve problems that appear to be very challenging, including problems of the kind that appear on the Math Olympiads.

CHAPTER 4: STRATEGIES, STRUCTURE AND STAMINA

I think I decided that I would never be a musician the day my kindergarten teacher made our class march in circles while she played on her upright piano. Sometimes she would interrupt the march to ask us to name a note when she hit it. I recall being mortified when I didn't get the right answer and running home in tears. In hindsight, I shouldn't have been so embarrassed. Only one in ten thousand people develop the ability to identify any note when it is played on a musical instrument. This ability is called "perfect pitch."

Until recently, psychologists believed that people can't develop perfect pitch unless they're born with a gift for it. But in 2014 the Japanese psychologist Ayako Sakakibara devised a remarkable experiment that put this belief to the test. Over several months, Sakakibara began by training twenty-four children between the ages of two and six to identify various chords played on the piano. As Anders Ericsson explains in *Peak*:

> *The chords were all major chords with three notes, such as a C-sharp major chord with the middle C and the E and G notes immediately above middle C. The children were given four or five short training sessions per day, each lasting just a few minutes,*

and each child continued training until he or she could identify all fourteen of the target chords that Sakakibara had selected. Some of the children completed the training in less than a year, while others took as long as a year and a half. Then, once a child had learned to identify the fourteen chords, Sakakibara tested that child to see if he or she could correctly name individual notes. After completing training every one of the children in the study had developed perfect pitch and could identify individual notes played on the piano.[1]

The outcome of this experiment is groundbreaking, not only because perfect pitch is such a rare and remarkable gift (thought to be possessed only by musical geniuses like Mozart), but because, until Sakakibara's experiment, almost everyone thought that either you were born with the gift or you weren't. The experiment inspires the question: what other extraordinary intellectual or physical gifts—including gifts that are uncommon and considered to be innate—can be unlocked through training programs like the one Sakakibara devised?

According to Ericsson, whose studies of highly talented individuals helped to create the science of expertise, people normally only develop exceptional abilities through dedicated practice. But Ericsson's studies showed that some forms of practice are more likely to produce these abilities than others.

In one form of practice, which Ericsson calls "purposeful" practice, a person sets clear goals and spends a great deal of time working in incrementally harder steps that push them just outside their comfort zone. People who practise purposefully will often see improvements in their skills, but their progress may be haphazard or limited because they don't have anyone to guide them.

The most effective kind of practice, according to Ericsson, is purposeful practice combined with feedback. In some fields, teachers and practitioners like Sakakibara have developed methods of practice that are extraordinarily efficient. When people practise purposefully with the guidance of an expert coach—as often happens in fields like competitive sports, chess and musical performance—they use a form of practice that Ericsson calls "deliberate" practice. The methods of deliberate practice that Ericsson outlines in *Peak* (for example, breaking practice into manageable chunks, getting constant feedback, raising the bar incrementally) are consistent with the methods of teaching that have been successfully implemented in JUMP. These methods can be used by anyone in any phase of life.

According to Ericsson, deliberate practice doesn't necessarily improve a person's ability to think, perform or solve problems outside the realm in which they are training. Professional batters who can hit balls travelling at ninety miles per hour don't typically have better reflexes than ordinary people. And grandmasters who can play a dozen opponents blindfolded are no better than the average person at recalling the positions of a collection of chess pieces that are randomly arranged on a board. Experts don't excel in their fields because they possess a broad range of mental or physical abilities that are superior to those of non-experts. They excel because, by using deliberate practice, they are able to develop better mental "representations" in their area of expertise.

When grandmasters see an arrangement of pieces that occurs in a game of chess (rather than a random arrangement), they see patterns that are invisible to non-experts. They can sense the strengths and weaknesses of various positions and know the outcomes of moves without playing out every possibility. They hold countless "chunks" of interrelated positions in their long-term memory and can recall these chunks in their entirety, without any effort, because the contents of each chunk are connected in meaningful ways. Out of a myriad of facts, rules, images and relationships they build mental representations that allow them to see deeper patterns and structures than non-experts see. Some chess players even speak of seeing "lines of force" on the board that guide them in their moves. Similarly, professional baseball players can see patterns in the trajectory of a ball moments after it has left the pitcher's hand—if they had to wait any longer to react, they would never hit the ball.

I argue that it's possible to design lessons in mathematics that allow learners of all ages to engage in the kinds of deliberate practice that produce experts in highly competitive fields such as chess, music or sports. To support this claim, I will show how teachers can systematically train students to solve the kinds of problems that appear on math competitions. Students who excel on math competitions are rare, just as people who have perfect pitch are rare. They tend to also excel in math at school and often go on to become mathematicians or scientists. As well, the strategies and approaches that students use to solve competition-style problems are exactly the ones I use in my research in math. So if it's possible to train people to solve the kinds of problems that appear on math competitions, then it should be possible to train them to do any kind of math. And if psychologists and cognitive scientists could demonstrate that people can be trained to solve these kinds of problems independently, it would help put to rest the idea that mathematical ability, like perfect pitch, is innate.

Structured Inquiry

One of my heroes in education is a cognitive scientist who also studied creative writing. Daniel Willingham's training in writing is evident in all of his talks and articles. Few cognitive scientists have his talent for translating research on learning and teaching into a form that is useful for educators. For two decades, he has worked tirelessly to dispel myths about how children learn and to empower teachers with new ideas and techniques that will improve their practice.

In his book *Why Don't Students Like School?*, Willingham makes a rather sobering claim about the brain: "Contrary to popular belief, the brain is not designed for thinking. It's designed to save you from having to think, because the brain is actually not very good at thinking."[2]

According to Willingham, thinking is "slow, effortful and unreliable." That's why the most important functions of the brain don't generally involve thought: they either are performed unconsciously (for example, when we perceive things using our senses) or rely on skills that have been committed to long-term memory and can be exercised without much mental effort (as when we drive a car after years of practice). Humans wouldn't have survived if the brain hadn't evolved to do much of its work automatically.

It's not hard to find examples from mathematics that support Willingham's claims. Computers perform calculations much more quickly and accurately than humans do, and they can see trends and patterns in data that are invisible to us. Historically, humans have been slow to discover even the most elementary mathematical concepts. The average Roman would have struggled to multiply a pair of two-digit numbers, because the Roman number system didn't have place values or a symbol for zero. Roman accountants and tax collectors must have performed millions of calculations, over the course of many centuries, to keep the business of the empire running, but none of them ever saw the advantages of using zero as a placeholder in their computations. Even after the empire collapsed, people in some parts of Italy were forbidden from using the symbol for zero because it was an "infidel number," invented by Persian mathematicians.

One reason people aren't very good at thinking is that they often lack the knowledge or skills they need to see any structure in the problems they encounter in various fields. People who are novices in a particular field will usually only see the surface details of the problems in that field, and they will have trouble perceiving the relationships between the various elements of a problem or knowing which relationships are unimportant and which are essential. It can take years of study and practice before they can see the deeper structure of those problems. In his book, Willingham cites a classic experiment that demonstrates this idea:

> *Physics novices (undergraduates who had taken one course) and physics experts (advanced graduate students and professors) were given twenty-four physics problems and asked to put them into categories. The novices created categories based on the objects in the problem: problems involving springs*

went into one category, problems using inclined planes went into another, and so on. The experts, in contrast, sorted the problems on the basis of the physical principles that were important in their solution. For example, all of the problems that relied on conservation of energy were put in one group whether they involved springs or planes.[3]

Even though humans aren't very good at thinking, we still do like to think, says Willingham, when the conditions are right. We enjoy solving problems and learning new things when we are reasonably confident that we will experience the feeling of satisfaction that comes when we succeed at overcoming a mental challenge.

Willingham's observations present a rather difficult dilemma for educators. On the one hand, people like to think when they believe that their efforts will be rewarded. On the other hand, people are not naturally good at thinking and will struggle to think productively if the cognitive conditions are not right. As well, our brains are easily overwhelmed when they have to absorb too much new information or apply too many new skills at the same time. And it can take many years of study and practice to develop the expertise required to see, without any assistance, the deep structure in many domains. That's why people will often avoid thinking. As Willingham says, "People like to solve problems but not to work on unsolvable problems. If school work is always just a bit too difficult for a student, it should be no surprise that she doesn't like school much."[4]

The method of structured inquiry is designed to resolve this dilemma. It enables the brain to think productively by striking a balance between independent and guided thought and between problems that are too hard or too easy for learners. In a lesson based on structured inquiry, students are allowed to enjoy the process of thinking, because the teacher doesn't spoon-feed them with answers or deprive them of the opportunity to explore and come up with ideas on their own. But the teacher also provides a great deal of guidance, scaffolding, immediate feedback and practice for students, because they know it would be foolish to expect students' brains to work the way expert brains work. To illustrate the method, I will start with a topic that is conceptually deep and has many practical applications, but that is rarely taught for full understanding or even proficiency. Then I will discuss how the method works with more challenging contest-level problems.

In North America, students learn to divide a multi-digit number by a one-digit number sometime between grades four and six. In some states, division questions in which students must divide a four-digit number by a one-digit number appear on state tests as early as grade four. In my training sessions, many elementary teachers have told me that only a minority of their students can perform the procedure (or "algorithm") for long division and even fewer students understand why the procedure works. I have found that the approach I describe below enables students to both discover the steps of the algorithm and understand the underlying concepts while learning to perform the algorithm proficiently.

I normally won't teach this lesson without giving students a brief review of the concept of division. I let them know that division statements are ambiguous (as I explained in chapter two) and that you need a context to know what the statement means. I tell students that when they are learning long division, it helps to start with a context or model of division that they are familiar with. I write the expression $3\overline{)72}$ on the board and tell students that, for the purposes of the lesson, they can interpret this notation to mean: 3 friends wish to share 7 dimes and 2 pennies (72 cents) as equally as possible. (Note: pennies are no longer used in Canada, but in my lessons I still use pennies or more abstract "one-cent coins," circles with a 1 on them, as they help students understand money and place value.)

Once students are familiar with the model of division that I want to use in the lesson, I ask them to draw a picture to show how they would divide the dimes among the friends. If students use a circle for each friend and an *X* for each dime, the diagram would look like this.

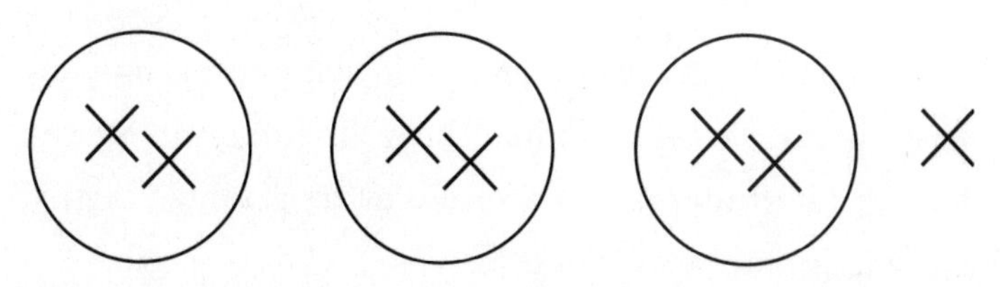

To make sure students understand the problem fully, I ask them to tell me the meaning of their diagram: each friend gets two dimes, and there is one dime left over.

I then tell students that if they happened to see an adult carrying out the first few steps of the long division algorithm, this is what they would see.

$$\begin{array}{r} 2 \\ 3\overline{)\,72} \\ -6 \\ \hline 1 \end{array}$$

I tell students that the adult would be unlikely to understand what they are doing when they perform "long division," and I challenge students to figure out what the steps in the algorithm mean by identifying where they see each number in their diagram. Students readily make the following connections between their diagram and the algorithm.

$$\begin{array}{rl} 2 & \cdots\cdots \text{ each friend got 2 dimes} \\ 3\overline{)\,72} & \\ -6 & \cdots\cdots \text{ 6 dimes were given away} \\ \hline 1 & \cdots\cdots \text{ there was 1 dime left over} \end{array}$$

I ask students to complete their diagram to show me how much money still has to be divided among the friends. If students use a circle to represent a penny, their diagram looks like this:

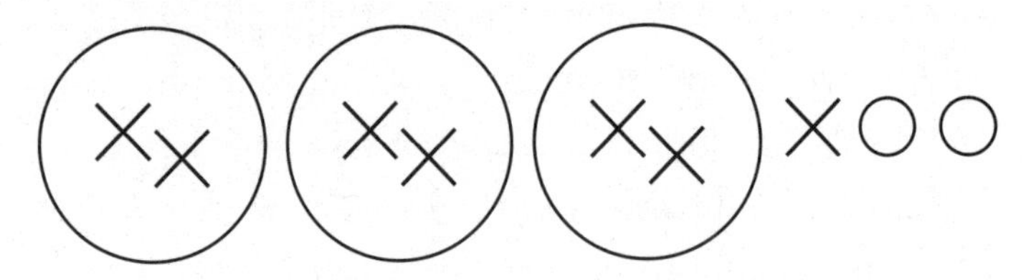

I invite three students to come to the front of the class so I can demonstrate how I would divide the remaining coins among the three friends. I give two students a penny each and one student a dime. The students always protest that my way of dividing up the coins isn't fair. They tell me they would exchange the dime for ten pennies and divide the twelve pennies among the friends. I inform students that this process of "regrouping" the tens (dimes) as ones (pennies) is actually a step in the long division algorithm. Most adults call this the "bring down" step, but very few understand it.

```
  2
3)72
 -6
 --
 12 .................. bring down the 2
```

When you "bring down" the number in the ones (pennies) column, you implicitly change the number in the tens (dimes) column into the smaller unit (pennies).

Then you combine all of your smaller units (to give twelve pennies altogether).

I then ask students to show me in their diagrams how they would divide the (twelve) remaining pennies among the friends. I also ask them to connect the numbers in their diagram with the remaining steps of the algorithm:

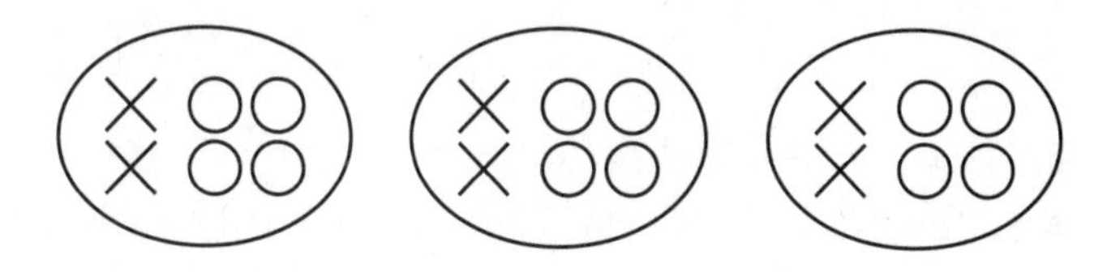

```
     24  ........ each friend received 4 pennies
   ----           (24 cents altogether)
 3 ) 72
    -6
    --
     12
     12  ........ 12 pennies were given out altogether
    ---
      0  ........ no pennies were left over
```

After each step in the lesson, I write several questions on the board so students can practise the step. I circulate around the class to see if any students need help understanding the questions or need more time to practise. Because students are likely to be deeply engaged in their work (they enjoy figuring out the answers to questions themselves, rather than being told the answers), and because the steps are simple and make sense, I normally have to help only a few students. As well, if students have mastered the previous steps, they have all the prior knowledge they need for the present step, so I can quickly resolve their problems and move on to the next step. Students who finish their work early can be assigned "bonus" questions (that are small conceptual variations on the work of the lesson), as I will explain in chapter six.

Every mathematical procedure that you learned (or didn't learn) at school can be reduced to steps or unravelled into conceptual threads that are as simple as the ones I outlined for division. This is true even for the advanced algebraic procedures that you confronted in high school. As well, a competent teacher can create a series of Socratic questions, exercises, activities and games that allow students to figure out why all of the steps of these procedures work, as in the example of long division above. That is why it is especially tragic that so many students fail to master and understand the mathematical procedures they are expected to learn at school or in training for their jobs.

Some teachers are reluctant to break instruction into manageable chunks or to unravel high-level concepts into threads that are easy to understand because they think that this kind of teaching is "rote." The term "rote" refers to a style of teaching where students are taught to blindly follow rules and procedures without any understanding of why those rules and procedures work. I hope it is clear from the lesson on long division that structured inquiry is not rote. In fact, research suggests that students who are led to explore and discover concepts through well-designed sequences of incrementally harder challenges will develop a much deeper understanding of math than students whose brains are constantly overwhelmed by lessons that introduce too many new concepts at the same time and don't give them enough time to practise and consolidate new skills and concepts.

Some teachers are averse to teaching in the way I have described because they think students should struggle in math classes so they learn to persevere. But, as Willingham points out, no one likes to struggle too much. And Carol Dweck made a similar point after watching the JUMP lesson on perimeter: she said the lesson incorporated growth mindset principles because the progression of exercises looked hard to the students, but weren't too hard. Of course different students will need different levels of challenge: in chapter six, I will discuss the research on motivation that has shed new light on this issue.

Even though I believe that teachers should learn to guide students' learning systematically, I don't advocate that they only teach this way. The JUMP lessons include exercises, games and activities that are much less structured than the long division lesson. And every

JUMP lesson ends with a set of "extension" questions that are more difficult than the questions covered in the lesson. As well, I recommend that teachers skip steps and give students more challenging or open-ended problems when they have developed the confidence and acquired the knowledge they need to struggle productively. My goal in teaching is always to help students become creative and independent thinkers who no longer depend on a teacher. But if teachers don't know how to unravel complex concepts into simple conceptual threads, then they are unlikely to help all learners reach a high level of achievement.

I've found that structured inquiry works particularly well with adults, who tend to learn math even more quickly than children when they are taught by this method. When adults decide that it is important to learn something, they can be more focused and determined than children. They also have more factual knowledge, more life experience and more exposure (formal or informal) to basic mathematics. That's why Darja Barr was able to cover so many topics with her nursing students in a one-week boot camp. A determined adult can even learn math without the help of a teacher, as Elisha Bonnis did when she worked her way through the online JUMP teachers' resources.

My claim about adult learners goes against the commonly held view that children should be able to learn almost anything more quickly than adults because their brains are more plastic. In her article "Children's Brains Are Different," neuroscientist Amy Bastian points out that when we talk about plasticity in young brains, we need to be more careful to define what we mean by heightened learning ability.

> *Children are super learners in the sense that they can learn to be more proficient than adults—that is they can gain fluency in a second language that is comparable to that of a native speaker. But this does not mean that every aspect of language learning is better. In fact, children learn a second language more slowly than adults. Similarly, it appears that younger children learn new movements at a slower rate compared to adults. Some work in this area has shown that this motor learning rate gradually improves (i.e., speeds up) through childhood and becomes adult-like by about age 12.*[5]

Bastian asks why, when children acquire individual motor skills less quickly than adults, they seem to learn some complex combinations of motor skills (like the ones involved in skiing or manipulating video consoles) more readily. She offers two explanations for this apparent paradox. First, children are far more variable in their movements than adults—they haven't settled into habitual patterns of movement and are more willing to experiment and contort their limbs and fingers into unfamiliar positions. Adults, on the other hand, "may be less willing to explore different movements and therefore tend to settle on a suboptimal motor pattern." And more importantly,

> *children may be more willing to undergo massive amounts of practice to learn motor skills. For example, when infants are learning to walk, they take about 2,400 steps and fall seventeen times for each hour of practice. This is an intense amount of activity; it means that infants cover the length of 7.7 American football fields per hour.*[6]

It is interesting to note that any advantages that children might have when it comes to learning to ski or play video games are mainly the product of habits of mind rather than superior motor skills. Similar factors may be at work whenever children appear to be able to learn math more quickly than adults. Just as children are more variable in their physical movements than adults, they are also more intellectually curious and tend to ask more original questions than adults—as this Twitter post by a perplexed parent shows:

> *My 5 year old son just asked, "What if we put a slice of turkey in the DVD player and it played a movie about the turkey's whole life," and none of the parenting books I've read have prepared me for this question.*[7]

As well as being more curious, children love repetition more than adults, and will happily practise math for hours if the challenges are made incrementally harder (as in a video game). But if adults can reawaken the sense of wonder they had as children and are willing to engage in regular practice, they can easily learn math.

One way I keep my own sense of wonder alive is by reading popularizations of science—books like *Chaos* and *The Information* by James Gleick or *The Fabric of the Cosmos* by Brian Greene. The more I read about science, the more motivated I am to learn math—so I can understand the mysteries that are revealed in these books more fully. And the better I become at math, the more I enjoy doing math. As Anders Ericsson observes in *Peak*, the sense of mastery people feel when they become better at something through practice eventually becomes its own reward.[8]

However, learners of all ages need to start at the right level and they need to be prepared to persevere. When physicist Joan Feynman was growing up in the 1930s, her mother told her she shouldn't attempt to learn physics, because it wasn't a proper vocation for women. Fortunately, her older brother Richard disagreed. He gave Joan a book on astronomy and advised her to learn the material using the following method: Read from the beginning, he said, until you don't understand anything. Then go back to the beginning and start again. Each time, he promised her, you'll go a little further.

Even though Joan faced a good deal of sexual discrimination on the road to becoming a physicist, she made important discoveries in geophysics and astrophysics and was awarded NASA's Exceptional Achievements Medal for her work. In the 1950s, when her brother was attending a conference near her home, she had an opportunity to return the favour he had done for her in her youth. By then Richard was regarded as one of the greatest physicists of the century. But during a talk at the conference—in which two young scientists presented a paper on some cutting-edge results in atomic theory—he momentarily lost his confidence. As he recalls:

> *I brought the paper home and said to her, "I can't understand these things that Lee and Yang are saying. It's all so complicated."*
>
> *"No," she said, "what you mean is not that you can't understand it, but that you didn't invent it. You didn't figure it out your own way, from hearing the clue. What you should do is imagine you're a student again, and take this paper upstairs, read every line of it, and check the equations. Then you'll understand it very easily."*
>
> *I took her advice, and checked through the whole thing, and found it to be very obvious and simple. I had been afraid to read it, thinking it was too difficult.*[9]

I find it very consoling that a scientist of Richard Feynman's stature could have doubts about his abilities and that, even in his prime, he sometimes had to learn his subject in the same methodical way that students do. Interestingly, Richard Feynman's IQ was only 123 and Joan's was 124 (she teased him about being more intelligent than he was). These are respectable scores, but not in the range of genius. This is one more piece of evidence that intelligence tests don't necessarily predict intellectual achievement or measure a person's full intellectual potential. According to sociologist Adam Grant: "When psychologists study history's most eminent and influential people, they discover that many of them weren't unusually gifted as children. And if you assemble a large group of child prodigies and follow them for their entire lives, you'll find that they don't outshine their less precocious peers from families of similar means."[10]

The Art and Science of Problem Solving

Most middle school students would have trouble solving the algebraic problem presented below. I have selected a type of problem that is frequently found on competitions like the Math Olympiads. Each letter in the sum stands for a unique one-digit number between 1 and 9; if a letter stands for a particular number, then no other letter can stand for that number. To solve the problem, you have to determine what number each letter stands for. But you can't simply assign values to the letters at random—when you replace the letters with your numbers, the sum has to work.

```
   H O S T
 + H O S T
 ---------
 T H E M E
```

When I was in grade seven or eight, I would have found this problem hard. (You might try the question before you read on to see if it's hard for you.) To understand how it might be possible to train any person to solve these kinds of problems, it helps to consider why I now find the problem easy. Because I am a trained mathematician, my brain has developed a number of capacities and habits of mind that allow me to solve the kinds of problems that appear on middle school competitions. If I could somehow become a thirteen-year-old again, while retaining my expertise, my peers and teachers would undoubtedly think I was a mathematical genius. That's because there are three key qualities that distinguish my brain from the brain of a person who hasn't had any training. To begin with, my brain is now able to employ a variety of strategies that I learned over the years that are extremely effective for problem solving. I also know much more about mathematics and mathematical structure than a young teenager could expect to learn. And I have developed a level of confidence in my abilities that motivates me to persevere much longer than a typical thirteen-year-old. In my research, I am still happily working on one research problem whose solution has eluded me for four years.

To show you how these three qualities (strategies, structure and stamina) play out in problem solving, I will walk through the steps I would follow. This will also give you an idea of how a mathematician's "mental representations" work.

One of the strategies I use most often in my research is the method of looking for patterns or violations of patterns. When I look at the letter problem, I immediately notice that every letter in the bottom row has two letters directly above it, except for the letter *T*. The *T*, at the left end of the bottom row, seems to stick out like a sore thumb, so I will start by trying to figure out what number it stands for.

In addition to using strategies, I will often rely on my knowledge of the structure of various mathematical operations (like addition or multiplication) or entities (like prime numbers or geometric shapes) to eliminate possible solutions to a problem. Because I understand the structure of addition, I know immediately, when I look at the *T*, what number it has to be. If you add 2 one-digit numbers, the greatest number you ever "regroup" or carry into the next column is 1. For example, when you add 9 plus 9, the sum is 18, so you will only carry a 1. This happens even if you add a string of nines, as shown below.

$$\begin{array}{r} 9 \\ +\ 9 \\ \hline 1\,8 \end{array} \qquad \begin{array}{r} 1 \\ 9\,9 \\ +\ 9\,9 \\ \hline 1\,9\,8 \end{array} \qquad \begin{array}{r} 1\,1 \\ 9\,9\,9 \\ +\ 9\,9\,9 \\ \hline 1\,9\,9\,8 \end{array}$$

So I know, without thinking, that the *T* in the letter problem has to be a 1. Now I can deduce that *E* is 2, because I also happen to know (from years of training) that 1 plus 1 equals 2!

$$\begin{array}{r} H\ O\ S\ 1 \\ +\ H\ O\ S\ 1 \\ \hline 1\ H\ 2\ M\ 2 \end{array}$$

Next I use my knowledge of the properties of odd and even numbers to eliminate some possibilities. When you add a number to itself, the result is always even. And when you add 1 to an even number, the result is always an odd number. So now I can deduce that *S* cannot be greater than 4. Otherwise, when I add *S* plus *S*, I would have to carry a 1 into the next column. Then, when I add this 1 to the sum of *O* plus *O*, I would get an odd number. But this possibility is ruled out because there is a 2 under the *O*. No odd number can end in 2.

I can now find the value of *O*. As I explained above, I know that a 1 is not being carried into the *O* column from the *S* column. So to find *O*, I simply need to find a one-digit number that will give me a 2 when I add it to itself. *O* can't be 1, because *T* is 1. That leaves only one possibility: *O* must be 6.

```
      1
      H 6 S 1
    + H 6 S 1
    ---------
    1 H 2 M 2
```

Another strategy that I frequently use in problem solving is called "guessing and checking." When I use this strategy, my goal isn't to hit on the answer to a problem by making a random guess. I try to make educated guesses, and these will sometimes lead me to the answer or help me understand the rules and constraints of the problem better.

I already know that *H* plus *H* must be greater than 9 because I need to carry a 1 to get *T*. That means *H* must be greater than 4. So I only have to guess and check a few numbers (5, 7, 8 and 9) to find it. Because I am confident about my abilities and have developed a good deal of stamina, I won't give up until I have solved the problem. So I guess and check until I find the only number that works, which is 9.

```
      1
      9 6 S 1
    + 9 6 S 1
    ---------
    1 9 2 M 2
```

I can also find the value of S by guessing and checking and using elimination. Recall that S is less than 5. S can't be 1 or 2 because these numbers are already used. S also can't be 3 because 3 plus 3 is 6, and 6 has also been used. So S must be 4, and M must be 8.

$$\begin{array}{r} {}^{1} \\ 9641 \\ +\ 9641 \\ \hline 19282 \end{array}$$

If you tried the problem on your own, you may have solved it in a different way. You might even have solved it in a better way. Because I once struggled in math, I always have a slight residual fear that someone will point out that I solved a particular problem in a very inelegant or inefficient way, or that I made a mistake. I have to constantly remind myself that although I sometimes make mistakes and don't always see the best solution to a problem, I have still managed to do some original work in math.

In my approach to the problem, I hope you can see the constant interplay between my strategies (which include looking for patterns, using logic to eliminate possibilities, and guessing and checking), my knowledge of structure and my stamina. Strategies and knowledge of structure can be learned. And, as I will explain in chapter six, new research in behavioural science suggests that stamina, or the ability to persevere and be deeply engaged in one's work, can also be learned.

In chess, one of the most effective ways to do deliberate practice is to solve chess problems. These problems are like mini chess games in which only a few pieces are positioned so that there are a limited number of moves that need to be tested to find the solution. Here is an example of a simple chess problem in which the black player must find a way to checkmate the white player in one move.

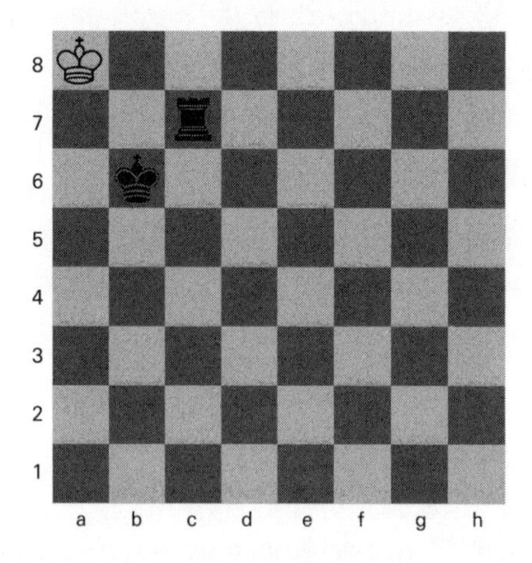

To solve the problem, a player needs to reduce the search for the best next move to a number of cases. In each case, the player mentally makes a move and checks what could happen next. This is equivalent to formulating a series of conditional statements: "If I do this, then this could happen." For example, in the problem shown in the figure above, the black player might think, "If I move my king to position A6, the white king can still escape to position B8." In real games, the chain of conditional statements can quickly become extremely complex: "If I move my pawn beside the king, the bishop could take it. If the bishop takes my pawn then my knight could take the bishop. If my knight takes the bishop then my opponent's king will have to move back one space diagonally to the left because moving my knight has exposed their king to my queen and there is no other place the king can go . . ."

In chess, it's easy to make a single move mentally. But players need to have a system for checking every possible move so they don't miss any possibilities. They also need to think through the implications of every case. In a more complex problem, there might be a very large number of cases. Strong players know how to see patterns and use their knowledge of good and bad positions to eliminate possibilities. With practice, they begin to see, without any mental effort, which cases are worth exploring and which aren't. If they didn't develop this ability, they would quickly become overwhelmed by how many possibilities are generated in a game.

If you look at the process I went through to solve the letter problem, you can see that it is similar to the process a chess player goes through to solve a chess problem. For example, to find the value of a particular letter, I sometimes broke the problem into cases. I formulated a conditional statement and then carefully analyzed the implications of the statement. When I was trying to find *O*, I thought, "If *S* is greater than 4, then I will have to carry a 1 into the *O* column. But that will give me an odd number under the *O*."

As I was analyzing each case, I was able to take many mental shortcuts. I knew without thinking that when I add a number to itself and then add 1, I will get an odd number. So I didn't have to check all combinations of one-digit numbers to know that I would get an odd number under the *O* if I carried a 1 into that column. As well, I was able to greatly reduce the number of possible solutions to the problem the moment I looked at it, because I saw that *T* had to be 1 and that *E* had to be 2. If I'd tried to randomly guess which numbers were represented by the letters, the odds are that I would have had to check thousands of possibilities before I hit on a combination that worked.

Books that are written to help people learn chess contain many sequences of chess problems like the one shown above. These sequences of problems are well scaffolded: they typically start with only three or four pieces but eventually add more pieces in more complex arrangements. By working through the problems, students gradually and systematically develop the mental representations they need to solve a wide range of chess problems.

Several years ago the JUMP writers and I set out to write a series of problem-solving lessons in math that mirror the approach in chess books. Although I can't present these lessons in any depth here, they are available online. I will sketch several stages of an approach that eventually enables students to solve letter problems like the one I just presented.

Here is an example of how I introduce younger students to the idea that a letter or symbol can stand for an unknown number. I ask students to close their eyes and I place a paper bag containing two blocks on a table. I also place three blocks on the table beside the bag and I place five blocks on another table. After I ask students to open their eyes, I tell them there are the same number of blocks on each table and I ask them to say how many blocks are in the bag. I also encourage them to explain their answer. Some students will say they found the answer by subtracting two from five because they know there are five blocks on the table with the bag and they can see three blocks outside the bag. Others count up from three, keeping count of how many times they counted, until they reach five.

I repeat this exercise with different numbers of blocks. Eventually I create problems using two or three bags with an equal number of blocks in each bag. For example, on one table I might place three blocks and two bags that each contain two blocks, and on the other table I might put seven blocks.

After students understand the game, I tell them I'm going to draw pictures to represent the problems, using a square to represent each bag and circles to represent the blocks (because they are easy to draw). I draw the following picture on the board and ask a volunteer to draw the missing blocks in the bag so there are the same number of blocks on each table. The following illustration shows how a teacher can gradually make the representation of the bag and block problem more abstract.

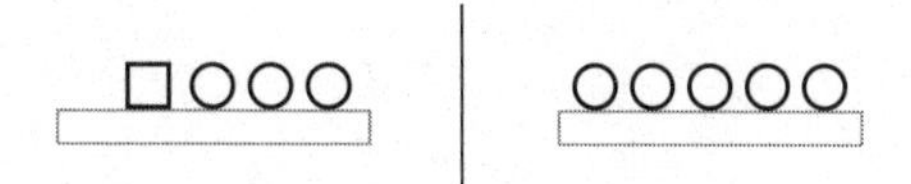

Gradually I make my picture more abstract.

□ ○○○ | ○○○○○

□ + 3 | 5

$\square + 3 = 5$

$\times + 3 = 5$

Once students understand that a letter can stand for a number and they have had some practice solving for unknowns, they can be introduced to more challenging problems. To help students learn to solve letter problems like the one I described above, JUMP writer Sindi Sabourin and I created a series of problems that lead from simple exercises with only a few digits (where numbers are mixed with letters) up to full contest-style problems like the HOST plus HOST problem we worked through above.

Here is a set of problems from the middle of the lesson that helps students learn to identify when there is regrouping or carrying involved in a sum.

B B	A B	A A	A A	A A	A B	A B
+ B	+ B	+ A	+ A	+ A	+ A B	+ A B
A 4	B 8	A 6	B 6	B 0	7 8	4 2

Bonus Questions

A A B	A A B	A A B	A B	A B
+ A B	+ A B	+ A B	+ A B	+ A B
A 7 8	B 7 8	C 4 6	B B C	B C C

If you work your way through this sequence of mini-problems, you should find that you are beginning to develop some strategies and mental shortcuts, like the ones I employed in the initial word problem above, that you could apply in more complex problems. (Remember that *A* and *B* have to be different numbers.) The set of questions I've given here is only a small subset of questions that lead all the way up to letter problems of the type I introduced above. If you would like to read the complete lesson, you can find it in the grade four US Teachers Guide on the jumpmath.org website.

Although letter problems are highly artificial and do not correspond to real-world situations, they are effective tools for training people to think mathematically.

The strategies I demonstrated in my analysis above are the same ones I use to solve problems and create proofs in my research. And the strategies that you learn in math can be applied to solve problems in any field.

In part two of this book, I will give other examples to show how mental representations (including visual representations) play a role in mathematical thinking. I will explain more fully how and why the method of structured inquiry can help learners develop mental representations and become expert problem solvers. I will also present seven principles of instruction—domain-specific knowledge, scaffolding, mastery, structure, variation, analogy and abstraction—that cognitive scientists have shown to be effective for learners of all ages and that are essential components of structured inquiry.

PART TWO:
PUTTING RESEARCH INTO PRACTICE

CHAPTER 5: THE SCIENCE OF LEARNING

There aren't many teachers who can say they learned the skills they needed to teach unruly classes of hormone-driven high school students while running nightclubs in Soho. But that is exactly where Tom Bennett first began to develop an approach to managing people that would qualify him to become an adviser on student behaviour to the UK Department of Education and earn him the title "Behaviour Czar" from the British media.

After graduating from high school, Bennett worked in London nightclubs for eight years before returning to university to do a degree in religion and history. He eventually landed a job as a high school history teacher and immediately set to work applying his expertise in the psychology of crowds in his classes.

In his school, Bennett saw first-hand how frequently teachers are asked to adopt methods of teaching that aren't backed by strong evidence. He watched a succession of fads sweep through the British educational system, and he fought to resist the pressure he and his peers often felt to adopt practices that sounded good in theory but that didn't work in the classroom. One night, in a state of frustration, he sent out a tweet that sparked a movement among teachers in England. And that movement soon spread around the world.

In his tweet, Bennett announced that he was putting together a conference on educational research (which he eventually named researchED) and asked if anyone was willing to help. Four hours later, he had received two hundred offers of help, moral support, venues and volunteer speakers. According to Bennett: "I didn't build researchED . . . it wanted to be built. It built itself. I just ran with it."

The first conference, held in 2013, attracted over five hundred teachers who were as frustrated as Bennett was about the lack of evidence in education and who were willing to dedicate a great deal of energy to the cause. As Bennett recalls:

> *It was genuinely moving, people offered their time and skills for nothing, without hesitation. From the logo design, to the name, to the people making up the name badges on the day, we were propelled by an army of the willing and able. I have never witnessed such organised, coherent, yet spontaneous kindness in my life.*[1]

The conferences quickly spread from London to venues across Britain and then to multiple cities on four continents. Many leading cognitive scientists and educational researchers have waived their hefty fees to speak at researchED conferences. And all of the conferences have been organized by volunteers.

The rapid growth of conferences like researchED has revealed a growing willingness among teachers to use rigorous research to improve their practice. In the past decade, researchers like Carol Dweck and Daniel Willingham have become sought-after keynote speakers at teachers' conferences. The popular journal *American Educator* recently published an article on teaching that included a long list of references to journals of psychology and cognitive science. The article lists ten principles of effective instruction that are well supported by research and that are embodied in the method of structured inquiry:

1. Begin a lesson with a short review of previous lessons.
2. Present new material in small steps with sufficient practice after each step.
3. Ask a large number of questions and check the responses of all students.
4. Provide models.
5. Guide student practice.
6. Check for student understanding.
7. Obtain a high success rate.
8. Provide scaffolds for difficult tasks.
9. Require and monitor independent practice.
10. Engage students in weekly and monthly review.[2]

When teachers fail to apply these principles in their lessons, it is usually not because they lack the skills they need to be effective teachers or because they don't want to help every one of their students succeed. It is usually because they are not aware of the research on how students learn or because they have been persuaded or forced, by educational consultants and administrators, to select resources and methods of instruction that are inherently inefficient. But more and more teachers are starting to learn about the research and to demand that they be allowed to use resources and methods of instruction that are evidence-based.

This is an exciting time for education. I believe our schools and workplaces will make huge strides in eradicating intellectual poverty in the next decade, because, as conferences like researchED have shown, the spread of scientific knowledge is unstoppable. In the next three chapters, I will discuss some of the general trends in research that will eventually transform our society by empowering every person to develop their full intellectual potential.

The Power of Knowledge

In 1988, psychologists Donna Recht and Lauren Leslie performed a classic experiment in which a group of middle school students who were considered to be poor readers found it much easier to comprehend a story about baseball than a group of their peers who were supposed to be strong readers:

One group of twelve-year-olds was academically knowledgeable and scored well on reading tests, but knew little about baseball. Another group didn't know much in the academic way, and therefore scored poorly on reading tests, but knew a lot about baseball. On this particular task, the sports fans proved to be better readers, and illustrated the general principle: when a topic is familiar, "poor" readers become "good" readers: moreover, when a topic is unfamiliar, normally better readers lose their advantage.[3]

Some educators believe that, in the age of the internet, teachers should focus on teaching students reading strategies, critical thinking skills and grammatical rules, rather than spending too much time building their vocabularies and content knowledge (since students can look up word meanings and facts online). But a growing body of research has shown that a person's capacity to understand a given passage of text depends as much on their knowledge of the subject matter as it does on their understanding of sentence structure or their ability to apply reading strategies. That's because the grammatical structure of a typical sentence can support many interpretations.

To show how much prior knowledge is required to interpret even the most common platitude, cognitive scientist Daniel Willingham suggests five different readings for the sentence "Time flies like an arrow."

Here is one reading of this sentence: if there was such a thing as a "time fly" (just as there are "fruit flies"), and if arrows were a type of food, then the sentence could be read as a statement about the culinary preferences of a certain type of insect.[4]

Psychologists have shown that content knowledge also plays a more important role in reasoning and problem solving than most educators realize. In *Why Knowledge Matters*, E. D. Hirsch argues: "Domain knowledge facilitates problem solving in any domain—hence the best way to teach 'problem solving skills' is to offer a broad education."[5] And the *Cambridge Handbook of Expertise and Expert Performance* claims: "Research clearly rejects the classical view of human cognition in which general abilities such as learning, reasoning, problem solving and concept formation correspond to capacities and abilities that can be studied independently of the content domain."[6]

To see how domain-specific knowledge plays a role in learning math, let's revisit the rule for division by a fraction that we examined in chapter three. If you recall, I helped you find the rule by providing pictures of a piece of chocolate cut into halves, then thirds, then hundredths. In each case, I asked you to predict the total number of fractional parts in the whole chocolate bar. If you had never been exposed to the concepts of multiplication or addition, then the only way you could find the number of parts would be to subdivide every piece in the chocolate bar and then count the total number of pieces. This process would be very time-consuming when the divisor is $\frac{1}{100}$. And while you were cutting up the chocolate bar into smaller and smaller pieces, you probably wouldn't see any deeper structure in my examples or discover a general rule that you could use to solve related problems quickly. You certainly wouldn't see that you can find the number of fractional parts in the whole chocolate bar by multiplying the denominator of the fraction by the number of pieces in the whole.

Some educators believe that it is no longer necessary to spend much time teaching kids mathematical facts or allowing them to practise calculations because (as with the argument about reading) they can look up anything they need to know on a computer. Instead, these educators argue, we should teach students how to find and analyze facts. In particular, we should avoid teaching them basic facts, like times tables, because these can only be taught through rote drills that take the joy out of mathematics and stifle a student's natural creativity. We should also avoid forcing students to practise basic skills, like adding and multiplying, because these things can be done by a calculator and this work will prevent them from developing their own methods. While there is some truth to this argument (children do need to learn how to find and analyze facts, and we should avoid boring them and stifling their creativity), it also has some serious problems, even apart from the fact that domain-specific knowledge is essential for understanding mathematical texts.

One reason students need basic knowledge in math is that human working memory, the temporary mental scratchpad we use so heavily in solving problems we haven't seen before, is very limited. On average it holds the equivalent of about seven numbers at a time, a limit that the demands of a complex problem can easily exceed if the problem requires knowledge a student does not yet have. Students who have not committed basic skills and facts to long-term memory have very little mental capacity left over to make inferences, integrate knowledge and reorganize information, so they struggle with complex problems. As well, people who don't know basic facts, like times tables, have trouble seeing patterns and connections between numbers or understanding rules or making predictions and estimates, as the example of division by a fraction shows.

According to Herb Simon, who was Anders Ericsson's thesis adviser (and a Nobel Prize winner in economics), several decades of studies in cognitive science have shown conclusively that students need practice to transfer the skills and facts that are required to solve complex problems to long-term memory. In his seminal article "Applications and Misapplications of Cognitive

Psychology to Mathematics," Simon says that there is no worse idea in education than the idea that practice is bad. He also laments the fact that educational theorists have made practice seem unnecessary or destructive by calling it "drill and kill." As he points out: "All evidence from the laboratory and from case studies of professionals indicates that real competence only comes with extensive practice. The instructional task is not to kill motivation by excessive drill, but to find tasks that provide practice while sustaining interest."[7]

Fortunately, practice can be made engaging for students when it is embedded in an incrementally harder series of challenges (as in a video game). To teach students basic facts, I often use exercises involving patterns. Students enjoy discovering the patterns, and the patterns help them recall the facts. For example, to help students remember the six times table, I write the following expressions on the board:

$2 \times 6 = 12$

$4 \times 6 = 24$

$6 \times 6 = 36$

$8 \times 6 = 48$

Students usually see many patterns in this array. For example, the numbers in the first and last columns are identical. This means that when you multiply 6 by an even number, the ones digit of the answer is the same as the number you are multiplying by. There is also an interesting relationship between the ones and the tens digit in each product: the tens digit is always half the ones digit (in 12, 1 is half of 2; in 24, 2 is half of 4; and so on). Once students have seen these patterns, they don't need to memorize these four entries of the six times table. In general, the practice students need to master basic skills and remember basic facts can be made more engaging through exercises of this kind.

Some of the most effective and practical educational innovations that have emerged from the new research on learning involve memory. When the brain learns new concepts or skills, it develops new networks of neural connections that encode that learning. But if the brain doesn't also build neural pathways to retrieve that learning, then the learning is lost. Cognitive scientists have found that people remember much more of what they learn if they employ simple strategies to develop and reinforce these pathways. But these strategies are not the ones people typically use when they are studying.

One popular method of studying involves rereading the material you want to learn over and over (preferably with a yellow highlighter in hand). But a slew of recent experiments have shown "self-testing" is more effective than simple review. Methods of self-testing might involve using flash cards, answering questions in a textbook or devising your own quizzes. In the article "What Works, What Doesn't," cognitive scientist John

Dunlosky and his colleagues summarize the results of recent research on self-testing or "retrieval practice":

> *In one study, undergraduates were asked to memorize word pairs, half of which were then included on a recall test. One week later the students remembered 35 percent of the word pairs they had been tested on, compared with only 4 percent of those they had not. In another demonstration, undergraduates were presented with Swahili-English word pairs, followed by either practice testing or review. Recall for items they had been repeatedly tested on was 80 percent, compared with only 36 percent for items they had restudied. One theory is that practice testing triggers a mental word search of long-term memory that activates related information, forming multiple memory pathways that make the information easier to access.*[8]

Another common method of study involves "massed practice," which means cramming or drilling the same material over and over in a short space of time. Research has shown that spreading out the same amount of study time over a longer period is more effective. As well, students remember more if they alternate between topics when they study, rather than studying in blocks (finishing the review of one topic before moving on to the next). In one study, college students learned to compute the volumes of four different geometric shapes. One group of students finished all the problems for one shape before moving on. Another group switched from one problem type to another as they worked, so they had an opportunity to repeatedly practise selecting the appropriate method for each problem. When the students were tested a week later, the group that used "interleaved" practice (switching between problems) rather than "blocked" practice were 43 percent more accurate in computing the volumes.[9]

The most recent research on memory and practice suggests that, through systematic practice and the retention of domain-specific knowledge, we have the power to become better problem solvers at school and in the world.

The Power of Scaffolding

In the past two decades, most schools in North America have adopted some kind of discovery- or reform-based math program, in which students are supposed to figure out concepts by themselves rather than being taught them explicitly. Discovery-based lessons tend to focus less on problems that can be solved by following a general rule, procedure or formula (such as "Find the perimeter of a rectangle 5 metres long and 4 metres wide") and more on complex problems based on real-world examples that can be tackled in more than one way or have more than one solution (such as "How could you estimate the area of this puddle?"). Instead of memorizing facts and learning standard algorithms such as long division, students learn math primarily by exploring concepts and developing their own methods of calculation, mostly through hands-on activities with concrete materials.

Although I agree with many of the aims and methods of the discovery approach, a growing body of research suggests that some of its elements have significant drawbacks.

Just as Jennifer Kaminski's work (introduced in chapter three) has shown that teachers should avoid selecting math resources that distract students with overly detailed or extraneous visual representations, research has also shown that teachers should avoid overwhelming students with too much new information or too many new cognitive demands at once. For example, Herb Simon has observed that when a large number of basic competencies are required to solve a complex problem, a learner who hasn't developed these competencies can easily be overwhelmed by the processing demands. But when component concepts are isolated and mastered in manageable chunks, learning occurs more efficiently.

Because of their hefty cognitive load, lessons based on pure discovery do not work as well as those in which a teacher helps a student navigate the complexities of a problem by providing feedback, working through examples and offering other guidance. According to a

2006 article by psychologist Paul Kirschner of the Open University of the Netherlands and his colleagues: "Empirical evidence collected over the past half-century consistently indicates that minimally guided instruction is less effective and less efficient than instructional approaches that place a strong emphasis on guidance of the learning process."[10]

In a 2011 meta-analysis (quantitative review) of 164 studies of discovery-based learning, psychologist Louis Alfieri of the City University of New York and his colleagues concluded: "Unassisted discovery does not benefit learners, whereas feedback, worked examples, scaffolding and elicited explanations do."[11]

In education, the term "scaffolding" means breaking learning into chunks and providing relevant examples and practice to help students tackle each chunk. In a properly scaffolded lesson, concepts are introduced in a logical progression, with one idea leading naturally to the next, and students are given feedback at each level to ensure that they can move on to the next one. Just as the scaffolding on a building allows a construction worker to climb safely and steadily to the top, the scaffolding in a lesson helps a student move upward to higher levels of learning.

A team of educational researchers from the University of Calgary, led by Brent Davis, have, since 2012, been observing and filming teachers while they deliver lessons based on the principles discussed in this book (with scaffolding, continuous assessment and so on).[12] Early in the study they tracked the results of two teachers who appeared to be using the lesson plans with the same degree of commitment. After a year, the scores of the students of one of the teachers had increased an average of 20 percentage points on the CTBS test (which measures performance in computation, conceptual understanding and problem solving), while the scores of the other teacher's students had not increased at all. When the researchers looked closely at the videotapes of the teachers in action, they saw some significant differences in the way they delivered their lessons.

The second teacher (whose students' performance didn't improve) would sometimes skip essential steps in the scaffolding of an exploration or activity, while the first had a better sense of what parts of the lesson could be safely left out and what parts couldn't. The second teacher would only assess her students occasionally and would not change her lesson plan on the basis of the assessment, while the first teacher was aware of what her students knew from moment to moment: she would frequently ask students to discuss their work or to hold up their answers on individual white boards and she would circulate around the class to watch students working. She would also change her lesson—by reteaching material or slowing down or speeding up—based on what the assessments revealed. The second teacher would rarely give bonus questions, while the first would regularly assign the "extension" questions from the JUMP lesson plans or create extra bonus questions to generate excitement. As the researchers gathered more data on the teachers in the study, they found that teachers who consistently used scaffolding, continuous assessment and incrementally harder challenges outperformed teachers who didn't.

In the fall of 2019, the researchers are releasing a free online course on the principles of instruction and lesson design that produced positive results in the study.

The Power of Mastery

It is not a new idea that almost every student can master concepts when they are provided with the right supports—including rigorous scaffolding and continuous feedback. In North America, the idea was first developed by several American educators in the 1920s and was revived in the late 1960s by educational psychologist Benjamin Bloom.

Bloom observed that when teachers grade students by using a bell curve to randomly distribute their marks on an exam, they implicitly teach with the expectation that many students will not be successful in learning the material covered. But if teachers take measures to ensure that all of their students master that material, then the achievement curve should be dramatically skewed toward high levels of achievement that will not fall on a bell curve.

In the 1980s, Bloom conducted a series of studies in which students were given tutorials based on "mastery" methods—where they could learn at their own pace while receiving continuous assessment with rigorous scaffolding and feedback. These students consistently showed levels of achievement that were much higher (on average, two standard deviations higher) than students taught with traditional methods. Bloom found that "about 90% of the tutored students . . . attained the level of summative achievement reached by only the highest 20%" of the control class.[13]

Bloom also applied the methods of mastery learning in regular classrooms and found that the methods produced results that were similar to (although less dramatic than) the results he observed in tutorials. According to psychologist Thomas Guskey, a large body of research conducted in the two decades after Bloom first developed his theories has borne Bloom out: "When compared with students in traditionally taught classes, students in well-implemented mastery learning classes consistently reach higher levels of achievement and develop greater confidence in their ability to learn and in themselves as learners."[14]

In spite of Bloom's striking results, interest in mastery learning began to wane in the 1980s, not long after he published his findings on mastery-based tutorials. Now many educators—particularly educational consultants who advise teachers—consider the approach to be old-fashioned or even damaging to students. The idea that all students should master the curriculum (or even its most basic elements) in mathematics at their grade level before they move on to the next grade is — as I can attest from my observations of hundreds of classrooms — no longer current in our school system. As early as grade five, the average teacher can expect a significant number of their students to be one to three grades behind when they enter their class in the fall. They can also expect — based on their experience using resources and methods of teaching that are not designed for mastery — that many students will be even further behind when they leave their class at the end of the year.

A grade seven teacher once told me a story that shows how foreign the idea of mastery is in our schools today. One day, when his students were measuring the lengths of various objects in centimetres and millimetres, he noticed that a student who generally did well in math was using her ruler in a rather curious way. Sometimes the girl would measure the length of an object correctly, by aligning the end of the object with the zero on her ruler, but other times, for no apparent reason, she would align the end of the object with the number one on her ruler, so that her measurement was off by one unit. When the teacher asked the girl why she was using the ruler in that way, she pointed to the letters *mm* and *cm* that were printed on her ruler. "When I want to measure an object in millimetres," she said, "I put the zero at the end of the object, because the letters *mm* are under the zero. And when I want to measure the object in centimetres, I put the one at the end of the object, because the letters *cm* are under the one."

Research in early childhood education has shown that, without guidance, some children don't naturally learn to use a ruler correctly—they need to be taught what the intervals on the ruler mean and how to align the ruler properly with the object they are measuring. The girl who measured millimetres one way and centimetres another had somehow managed to reach grade seven before anyone noticed that she didn't know how to use a ruler. In the same way, many students progress from grade to grade without learning the essential skills and facts they will need to understand higher levels of math. Many high school teachers have told me that a significant proportion of students entering grade nine have a limited understanding of concepts involving fractions, decimals, ratios, percentages, integers and simple algebra, and are reliant on calculators to perform even the simplest operations.

In my book *The End of Ignorance*, I described my work with Lisa and Matthew, two of the early JUMP students (when the program was still a tutoring club in my apartment) who both had severe learning disabilities. I still vividly remember the day when Lisa—a tall and painfully shy grade six student—sat down at my kitchen table for her first math lesson. Although I'd asked Lisa's principal to recommend students who were struggling in math for the program, I wasn't prepared for the challenges I would encounter teaching Lisa.

I'd planned to boost Lisa's confidence by teaching her to add fractions. I knew from previous experience as a tutor that children often struggle and develop anxieties in math when they first encounter fractions. Because my lesson would involve multiplication, I asked Lisa if there were any times tables she had trouble remembering, and she stared at me blankly. She had no idea what multiplication meant. Even the concept of counting by a number other than one was foreign to her. She was terrified by my questions and kept saying, when I mentioned the simplest concepts, "I don't understand." She also had trouble reading and told me she had never read a chapter book in her life. I found out later that she had recently tested at a grade one level and had been assessed as having a mild intellectual disability.

I had no idea where to start with Lisa, so I decided to see if she could learn the sequence of even numbers (2, 4, 6, 8 . . .) and eventually be able to multiply by two. Because she was so nervous and had trouble remembering the numbers, I told her I was certain she was smart enough to learn to multiply. I was afraid I might be giving her false praise, but my encouragement seemed to help her focus and she made more progress in the lesson than I expected.

I tutored Lisa once a week for three years. By the time she entered high school, I could see that my worries about the false praise were unfounded. In grade nine, she transferred out of the remedial stream in math, and in her second term, she skipped a year and enrolled in grade ten math. She occasionally failed a test, but more often scored in the C to A range. She was able to solve word problems and carry out complex operations on tests independently, and several times I watched her teach herself material out of a textbook. Her final mark in grade ten math was a C plus, but she was a year ahead. She had progressed from grade one to grade nine in only a hundred hours of lessons (fewer than she would have received in a year of school). If I'd had more time to prepare her, I'm sure she could have done better.

I began teaching Matthew – a boy with autism who faced even greater challenges than Lisa – after his doctor read about JUMP in the news and contacted me to ask if I would work with him. In grade four, Matthew had been diagnosed as being in the 0.1 percentile in mathematical ability, which means that on average he would test lower than 999 out of 1,000 individuals. He had been removed from regular lessons at school because he was so anxious about math that it would sometimes cause him to throw up in class.

I used the methods I describe in this book to teach Matthew. Happily, I can provide an update on his progress here. I tutored Matthew once every two weeks (on average) from 2003 until 2010. He also received occasional tutoring from one of the JUMP writers, Francisco Kibaldi. In 2010, when Matthew's doctor reassessed his abilities in math, he scored in the low average range (around the 20th percentile), which is a dramatic jump from the 0.1 percentile. By then, Matthew had become much more confident about his abilities and appeared to enjoy his math lessons. As with Lisa, I expect he could have gone further if I had been able to teach him every day (and if I was a better teacher).

Matthew and Lisa are extreme cases. Most of my students have been much easier to teach. But I feel lucky to have had an opportunity to work with them, as I learned a great deal about teaching and the potential of children in our lessons. They helped me see that even students with significant challenges can learn mathematics if a teacher uses mastery-based methods of teaching.

If I'd thought that Lisa's and Matthew's abilities were fixed and couldn't be improved by practice, I doubt I could have helped them improve as much as they did. And I also wouldn't have developed my own skills as a teacher, as I would have been less likely to try new approaches and learn from my mistakes. Now, when one of my students fails to understand something I am trying to teach them, I always assume that the fault lies in my lesson and not in their lack of ability. And when a student is stuck, I try to determine if something in my approach was confusing or difficult for them, or if their prior knowledge of a topic (which might be based on a misconception) is getting in the way.

It can take a great deal of patience, repetition and trial and error to help some students. I have worked with a few students who were so severely learning disabled that I didn't make much progress with them, but even these students enjoyed learning math and benefited from what they learned.

One of Mary Jane Moreau's students, a girl who scored in the 9th percentile on the TOMA at the beginning of grade five, was at least three years behind when she entered Moreau's class. In the first half of the year, the girl didn't always do exactly the same work as the other students and she didn't do as well on tests. But Moreau made sure the girl couldn't easily compare her rate of progress with that of other students, and she gave her special bonus questions to build her confidence. She also gave her extra time to practise the basic skills she needed to reach grade level. When the girl occasionally noticed that she was behind, Moreau would reassure her that if she made an effort, she would catch up. At the end of grade five, the girl scored in the 95th percentile on the TOMA, and a year later, she came within three points of earning distinction on the Pythagoras competition. This story shows the power of practice and the importance of letting students know (as Carol Dweck recommends) that even if they can't do something "yet," they will with effort.

Some proponents of mastery teaching think that the approach never caught on in schools, in spite of the empirical evidence supporting it, because it was too demanding for teachers. If teachers were expected to use mastery-based methods, they would need to find or create lessons that were more carefully scaffolded than the lessons found in traditional textbooks. They would also need to track every student's progress carefully and find time to reteach material that students failed to learn. Although these aspects of mastery learning are indeed challenging, many teachers have found that, after a year or two of practice with the JUMP lesson plans, they can internalize the scaffolding and deliver the lessons without having to do extra work.

While the demands of mastery-based teaching may have played a role in slowing the dissemination of Bloom's ideas, I believe there are deeper reasons why his approach was not more widely adopted in schools. In *The End of Ignorance*, I argued that educators often mistake the ends of education with the means to achieve those ends. I believe that this tendency is the main reason mastery learning never caught on.

In competitive chess, the goal of training is to become good at playing the game of chess. But players don't learn very efficiently if they simply play the game over and over. They advance more quickly by doing focused exercises based on incremental challenges. The goal of training in chess (playing the full, unrestricted game) and the means to achieve that goal (learning in artificially constrained chunks with lots of repetition) look very different.

Hoping to produce highly creative and ingenious "twenty-first-century learners"—who are driven by their insatiable curiosity to persevere in the face of the most difficult intellectual challenges—schools will continually give students batteries of "rich" problems that can only be solved by the few who already have the knowledge, skills and habits of mind that the problems are meant to teach. Because this approach to problem solving causes cognitive overload in most children, it has produced a great many students who are curious only about what they must learn to pass the next test and who demonstrate enormous creativity and resourcefulness in avoiding any real work involving math. If our goal was to produce students who have a good sense of humour, our methods of teaching would be well suited for that purpose. A number of websites have posted the work of students who found ingenious ways to turn their confusion about math into humour. When asked to "Find *x*," in a question where *x* appeared beside the hypotenuse of a right-angle triangle, one clever student drew an arrow pointing at the *x* and wrote the caption "Here it is."

The idea that children will become good at math only if they are given many opportunities to work on hard problems with little guidance or preparation has done enormous damage in our schools, particularly in neighbourhoods where parents can't afford to hire tutors to teach children the basic skills and concepts that they need to solve the problems. And unfortunately for disadvantaged students, our tendency to mistake the ends for the means runs very deep. I have heard many well-intentioned principals and teachers, in schools where only a small fraction of their students are at grade level, argue passionately for their students' right to engage, in every lesson, with problems that they are ill-equipped to solve.

I believe that all students should indeed have a right to tackle rich, interesting problems that test their ingenuity and stretch their capacity to persevere in the face of challenges. But the research shows that they are more likely to benefit from working on these types of problems if they are first allowed to master the skills and concepts they need to do this kind of work.

The Power of Structure

In most math textbooks today, there are many more words than numbers. The books are typically full of "word problems" that try to make math relevant by asking students to apply their mathematical knowledge in real-world contexts. On state and provincial exams, these are usually the problems that separate students who do well in math from students who don't.

Sometimes students struggle with word problems because the text is too hard for them. But even when this isn't the case, they may still have trouble seeing the mathematical structure that is buried under the words. Teachers often try to help students who struggle with word problems by giving them more word problems. This remedy can have the same effect as pouring gas on a fire—it reinforces a student's sense of failure and makes it harder for them to develop the confidence and ability to focus they will need to solve the problems.

In grades two and three, students sometimes struggle with word problems that involve a collection of things (or a "whole") that comprises two different kinds of things, or two "parts." If we are told that Bob has 4 marbles and Alice has 3 marbles, it's not hard to see that we can add 4 plus 3 to find how many marbles they have altogether. But if we are told how many more marbles one person has, the problem becomes more difficult. Some teachers like to tell students to look for key words in a problem, and that when they see the word "more," it means they need to add to find the answer. But this isn't always the case. If Bob has 6 marbles and Alice has 2 more marbles than Bob, we would add 6 plus 2 to find out how many marbles Alice has. But if Bob has 6 marbles and has 2 more marbles than Alice, we would subtract 2 from 6 to find out how many marbles Alice has.

When students are required to balance the cognitive demands of reading a series of word problems—where the vocabulary and context may change from problem to problem—with the demands of recognizing what problem type they are given, they can easily suffer from cognitive overload. The more elements of a problem teachers vary at one time, the more likely it is that they will leave students behind.

One way to address this problem is to allow students to practise finding solutions to various types of part-whole problems in a series of exercises where only the problem type varies but where the numbers are small and the language is minimal and doesn't change. Rather than being asked to read whole paragraphs—about animals, then cars, then vegetables—students might be given short phrases that are always about, for example, green and blue marbles. They might also be given the easiest problem type (where you are given the part and the part) first and then progress to harder types.

The question below shows the easiest problem types, when you are given the two parts and are asked to find the total and the difference.

The part and the part.

Shade boxes to show the number of marbles.

Then find the total and the difference.

5 green marbles

3 blue marbles

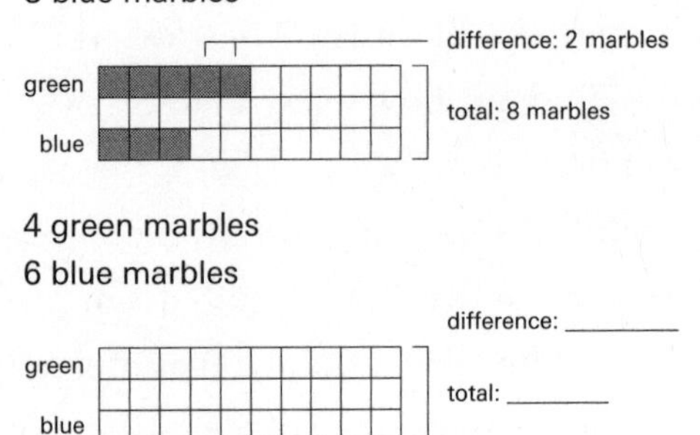

Students might also be allowed to practise solving each type as many times as they need to, so they understand one type before they are introduced to the next. The questions below show the other problem types.

The part and the whole

6 marbles altogether
2 green marbles

______ difference: ______

______ total: ______

The part and the difference, when you know the smaller part and you know how many more

3 green marbles
4 more blue marbles than green marbles

______ difference: ______

______ total: ______

The part and the difference, when you know the bigger part and how many fewer

8 green marbles
3 fewer blue marbles than green marbles

______ difference: ______

______ total: ______

Once students have mastered the various types of part-whole problem individually, they will still need to practise recognizing the various types when they are presented at random. Switching between different problem types is not an easy task for some children, even when they have mastered each type separately. If the problems are presented in full paragraphs, it can interfere with the acquisition of this skill. So the teacher could put the information for each problem into a table, as shown below.

	green marbles	blue marbles	total	difference
a.	3	5	8	2 more blue marbles than green
b.	2	9		
c.	4		6	
d.		2	7	
e.	6		10	
f.	3			1 more blue marble than green
g.		2		1 more green marble than blue
h.		4		1 more blue marble than green

For students who are capable of handling bigger numbers, the teacher could include in the tables some numbers that are larger than the number of squares in their grids. This will stretch students a little, by forcing them to either draw their own sketch or rely on their knowledge of numbers to find the answer mentally. The research on deliberate practice suggests that students learn most efficiently when they are continually pushed a little outside their comfort zone, but not too far.

Sometimes students who have been taken through this series of exercises will still panic when they see a word problem presented in a full paragraph. They will revert to guessing answers, even if they were able to solve the problems when they were presented with minimal language. To break this guessing reflex, one of the JUMP writers, Anna Klebanov, came up with an ingenious solution. As shown below, she put fragments of word problems (in which marbles were replaced by fish, so students could get used to changing contexts) on the left-hand side of the chart. Rather than asking students to write the answer to the problem, she asked them to fill in all of the missing pieces of information in the chart, then circle the answer. Students are forced to develop a complete (mental or physical) picture of the situation before they are allowed to answer the question. This stops the student from simply guessing.

	red	green	total	difference
a. Kate has 3 green fish and 4 red fish. How many fish does she have altogether?	4	3	(7)	1
b. Bill has 4 green fish and 6 red fish. How many fish does he have altogether?				
c. Mary has 8 green fish and 2 more green fish than red fish. How many fish does she have?				
d. Peter has 19 fish. He has 15 green fish. How many red fish does he have?				
e. Hanna has 8 green fish and 3 fewer red fish than green fish. How many fish does she have?				
f. Ken has 22 red fish and 33 green fish. How many more green fish does he have?				

After assigning this kind of exercise, the teacher can introduce problems with more text where the context varies from problem to problem.

The most challenging type of part-whole problem is one where the student is given the total and the difference. For example: "You have 20 marbles. You have 4 more green marbles than blue marbles. How many blue marbles do you have?" This is a perfect bonus question for students who are ready to be stretched a little further.

Only the most challenged student would encounter any real barriers or pitfalls in the progression of exercises I just outlined. When I teach in this way, I usually find that all of my students can move at roughly the same pace and no one is left behind. I can also cover material fairly quickly, because students are engaged and their brains aren't being overloaded. I always have a stock of incrementally harder bonus questions, so no one is bored. If students are ready to be stretched more, I can skip steps and let them struggle more. Although I haven't had the opportunity to test this series of exercises in a rigorous study, I predict that this approach would yield better results than an approach that involves giving students full word problems at the beginning of instruction.

When the language is stripped away from part-whole problems, you can see how easy the math is. In order to know everything in such problems, you only have to know the two parts, the total and the difference. Once you are able to form a mental representation of the situation (for example, by visualizing two bars representing the parts), you have plumbed the depths of the mathematics. There is no hidden mystery here that only the most brilliant people can understand. Fortunately, this is the case with all of the mathematics students are required to learn at school: it can all be taught in a series of steps that almost anyone can understand. That's because the underlying structure of the math is invariably simple and accessible to virtually anyone, as long as that structure isn't obscured by language or the learning made difficult by too many cognitive demands imposed at the same time.

When language is allowed to get in the way of math, children who speak English as a second language or who are weaker readers suffer the most. And when these students fall behind in math, the consequences are particularly tragic, because math is the subject where they could most easily shine and develop confidence. In learning math, they could also develop many abilities that transfer to other subjects, such as the ability to focus and stay on task, to see patterns in strings of symbols, to form mental representations of physical situations, to reason and recognize sound arguments and proofs, to apply strategies and so on. In North America, the campaign by privileged, well-educated people to make math instruction more and more dependent on reading ability (as evidenced by the density of text in textbooks) is not only unscientific; it is also an easily avoidable form of social injustice inflicted on disadvantaged students.

Later in this book, we will look at several other myths about learning, including the myth that individual "learning styles" are hard-wired into our brains and that students can only learn when they are taught according to their style, or the myth that children will naturally acquire mathematical concepts by playing with lots of concrete materials.

The strategies I've discussed in this chapter could dramatically influence the educational trajectory of the millions of children who struggle in math and the adults they become. When math is taught in ways that constantly overwhelm the brain, people learn to give up whenever they are confronted with a non-trivial mathematical problem. Some even develop a belief that all forms of knowledge that are based on numbers or logic (like statistics, the science of climate change or economic theory) are too arcane for ordinary people and are not to be trusted. I believe the recent rise of antiscientific political movements is tied to our failure to make math and science accessible to the majority of students.

When Jane Goodall was interviewed by *Scientific American* in 2010, she explained why she now spends so much of her time travelling the world giving talks, rather than continuing the research on chimpanzees that made her famous.

> *Young people everywhere need to realize what we do individually virtually every day does make a difference. If everybody begins thinking of the consequences of the little choices they make—what they eat, what they wear, what they buy, how they get from A to B—and acting accordingly, these millions of small changes will create the larger changes we must have if we care at all for our children. This is why I am on the road 300 days a year talking to groups of youths as well as adults, politicians and businesses—because I don't think we've got that much time left.*[15]

There are very few things I enjoy more than sitting at a desk doing mathematics. This peaceful and sedentary activity allows me to roam the universe and explore the hidden structure of the world without leaving my chair, with nothing more than a pencil and paper and my imagination. But, like Goodall, I don't spend as much time on my research as I would like, because I agree that we have very little time to solve our most pressing problems, including global warming.

Bertolt Brecht captured the sense of anxiety I feel about our age in his prescient poem "To Posterity," which he wrote in the 1930s:

Ah, what an age it is
When to speak of trees is almost a crime
For it is a kind of silence about injustice! . . .
They tell me: eat and drink. Be glad you have it!
But how can I eat and drink
When my food is snatched from the hungry
And my glass of water belongs to the thirsty? . . .

The old books tell us what wisdom is:
Avoid the strife of the world
Live out your little time
Fearing no one
Using no violence
Returning good for evil—
. . . I can do none of this:
Indeed I live in the dark ages![16]

On many fronts, especially in our efforts to protect the environment and reduce economic inequalities, we haven't progressed very far since Brecht wrote this poem—in part because we haven't made a concerted effort to eradicate intellectual poverty. I founded JUMP because I believe that one of the most efficient and cost-effective ways to improve our condition is to give people the intellectual tools they need to, as Goodall says, think about "the consequences of the little choices they make." People also need a sense of agency that will motivate them to tackle our most serious problems.

Fortunately, the research that I will discuss in the next two chapters gives me a good deal of hope. If people take the implications of this research to heart, they can develop the mindsets and creative resources they need to change things for the better.

One of my students, Ned, suffered from a severe attention deficit disorder. In our tutorials, he would sometimes drift into a kind of fugue state and sit staring into space, with no apparent awareness of what was happening around him. He had a gift for steering a conversation onto any topic other than the one I wanted him to focus on. He also had trouble remembering number facts and didn't know any times tables, even though he was in grade four. Because he had so much trouble paying attention, I had to work very hard to help him stay on task.

I knew that Ned had somehow managed to learn to read numbers in the hundreds, so in one of our lessons I told him I was going to give him a challenge: I would show him how to double large numbers mentally. I wrote the following:

million	thousand	
2 3 4	1 2 2	1 4 1

I covered all but the millions part of the number with my hand and asked him to read what he could see. He said, "Two hundred thirty-four" and then "million." I drew back my hand to reveal the thousand part, and he said, "One hundred twenty-two thousand." When I exposed the rest of the number, he said, "One hundred forty-one." As I had hoped, Ned became very excited about reading this enormous number and asked to read more. Soon he was reading numbers in the billions.

One of my goals in the lesson was to motivate Ned to remember some of his multiplication facts. So, after reviewing the meaning of multiplication, I made a list of the first four entries of the two times table and showed him how to double a large number by doubling each digit and writing the result under the digit. While he was happily doubling numbers, Ned memorized the list and soon no longer needed it. Because he was so engaged, he had practised and learned part of the two times table in several minutes without being aware of it.

I've done this same attention-building exercise in many classrooms. The exercise has an even greater impact when I do it with a group, rather than with an individual student, because the students in the group invariably want to have a turn to demonstrate, in front of their peers, that they can read big numbers. Once, when I did the lesson with a grade two class, I said, "I can't give you anything harder . . . I could go to trillions or quadrillions," and the students yelled, "Yes!" (You can watch part of this lesson in my TEXxCERN talk, The Mathematics of Learning.)

I've done similar lessons as early as kindergarten. I draw two dots on the board and ask a volunteer to join the dots. Then I move the dots farther apart and ask a second volunteer to join them. I do this several times until one of the dots is at one end of the board and the other dot is at the other end. By this time the children are usually so excited to come up and meet the challenge that they can barely sit still. Then I repeat the exercise, but I place one dot above the other, so the kids have to draw a vertical line to join them. Then I place the dots diagonally. The students seem to think that each variation is harder than the last and become more and more engaged. Eventually, I draw multiple dots, which I number, and ask students to join the dots in order, counting as they go. I even ask them to predict what letters or shapes they will make when they join the dots. I find this exercise helps kids remember the counting sequence and prepares them for writing their letters and numbers.

I've also done similar confidence-building lessons in grades eight and nine, where, for instance, I show students that within thirty minutes or so they can learn to solve rather complex-looking equations. The interesting thing about these lessons is that they tend to have the same impact in every classroom. Students respond enthusiastically to the incremental variations and to the fact that, in their eyes, the math appears to be challenging or beyond grade level. As well, all or almost all of the students, including those who have attention or behavioural problems, are able to focus and participate in the lesson.

More Alike Than Different*

Many people find it hard to believe that an entire class of children could become deeply engaged, in much the same way, in a lesson on math. The prevailing view in education and in our society is that children are unique individuals; that they learn in very different ways, at different paces, and are motivated to learn by very different things. According to this view, children even have different learning styles: some are kinesthetic learners (they learn best through movement), some are auditory learners (they learn best by hearing), and others are visual learners (they learn best by seeing). Children also have different intelligences: some might have a higher musical intelligence and others a higher mathematical intelligence.

Many educators believe that students learn better when the materials or methods the teacher uses for a lesson match their students' cognitive style. For example, a visual learner might learn a list of words more easily if they are shown a series of pictures depicting the words. This view of learning is called the theory of learning styles.

Over the past decade, many prominent cognitive scientists have written articles or books calling into question the theory of learning styles. In 2008, a team of cognitive psychologists were commissioned to conduct

---

* In *Why Don't Students Like School*, Daniel Willingham says that a guiding principle in education should be that "Students are more alike than different."

a literature review to determine whether the theory is supported by empirical evidence. The team found that most studies of learning styles were not well designed and were consequently not capable of testing the theory's validity. The few studies that were well designed produced either no evidence or negative evidence. Moreover, according to psychologists Henry Roediger and Mark McDaniel, the review "showed that it is more important that the mode of instruction match the nature of the subject being taught: visual instruction for geometry and geography, verbal instruction for poetry and so on. When instructional style matches the nature of the content, all learners learn better, regardless of their differing preference for how the material is taught."[1]

According to Daniel Willingham, 90 percent of teachers believe in the theory of learning styles.[2] One reason so many educators believe in a theory that is not supported by rigorous evidence may be that children are indisputably different. Some are better at remembering images and others are better at remembering sounds. Some love classical poetry and others prefer rap. No one can deny that children have different tastes and interests or different cognitive abilities and intelligences.

But cognitive scientists *don't* deny these things. They merely point out that even if children do have different interests and abilities, they won't necessarily learn better in the mode of instruction that they prefer. As Willingham puts it:

> *Mathematical concepts have to be learned mathematically, and skill in music won't help. Writing a poem about the arc that a golf club should take will not help your swing. These abilities are not completely insulated from one another, but they are separate enough that you can't take one skill you are good at and leverage it to bolster a weakness.*[3]

Although I have presented arguments against the theory of learning styles, I don't mean to suggest that teachers shouldn't sometimes draw on students' tastes or preferences to motivate them to learn. Some teachers, for example, have used rap music or art to help students develop an interest in math, by showing them that the subject can be cool or beautiful or relevant. However, once students are willing to pay attention to a lesson on math, much of the instruction will still have to be done in a mode that suits the content.

When I break challenges into manageable chunks and raise the level of difficulty incrementally, students become engaged in math, even when I don't do anything else to spark their interest. I don't have to feed them pizzas cut into fractional parts to motivate them to work on bonus questions on fractions. If the lesson is properly scaffolded, students are able to learn in any mode or with any representation that suits the content. I don't teach number lines because I think they will help visual learners—I teach them because they embody the abstract structure of many real-world situations and are powerful tools for problem solving. Fortunately, students don't need to be talented artists or have strong visual imaginations to sketch or visualize a simple number line. And this is true of most representations in math. In fact, research suggests that less detailed representations, which are accessible to any student, are often the most effective teaching tools.

Educational theories that emphasize the differences between students seem very benign. After all, what harm can come from regarding students as unique individuals with unique needs and interests? But I believe these theories may play a role in producing some of the differences between students that people think are natural. Because we are overly focused on children's differences in our schools, we have failed to pay attention to the research that shows their brains all work in roughly the same way and have roughly the same potential. By disregarding the principles of learning that apply to every brain—that emphasize, for example, the importance of scaffolding, feedback and practice—we have artificially imposed vastly different standards of academic achievement on our students.

I argue that our theories of learning have also produced schools where students demonstrate vastly different levels of academic engagement. Just as there are universal principles that determine how we learn, there are also universal principles that determine how we become motivated to learn. The kindergarten students who can't wait to join a pair of dots when they are placed at opposite ends of a blackboard or the grade eight students who love to show off their prowess with difficult-looking equations are motivated by the same drives that compel adults to set sail for uncharted seas, to push their bodies to new levels of endurance or to postulate new theories of the cosmos. If we wish to harness these basic human drives to help every student learn, we need to recast both our theories of motivation and our theories of achievement so they are founded on principles of equality rather than difference.

The Science of Motivation

In his book *Drive*, Daniel Pink draws on several decades of research in behavioural science to frame a novel theory of motivation that helps to explain why students become excited about learning math when I use the methods of instruction described above. The book has been praised by behavioural scientists and also has inspired many business leaders to rethink the way they motivate their employees. Pink argues that once people's basic financial and material needs are met, they are primarily motivated by three deep and enduring desires: they want to engage in activities that have a purpose or meaning; they want to achieve mastery in the things they learn and do; and they want to have a sense of autonomy, knowing that they are in control of their choices.[4] Let's look at the various ways in which these three drives can influence students' behaviour.

When teachers try to motivate their students, they often rely on "extrinsic" rewards. These rewards—which might include gold stars for good behaviour, high marks on a test, or medals for sports—are given to students to persuade them to work harder or perform better at school. These kinds of rewards are called extrinsic because they are not directly under the control of the person receiving the reward and also are not inherent in or a direct product of the activity for which the reward is given.

Research in psychology has shown that extrinsic rewards can sometimes have rather surprising unintended consequences. In one study, researchers identified a group of kindergarten children who loved to spend time drawing on their own, without any prompting from adults, just for the sheer pleasure of it. Half of the children in the study were given rewards for their drawing and the other half weren't. After several weeks of this treatment, the researchers found that the children who received the rewards (tragically) spent less time drawing and were less engaged in it, while the children who didn't receive rewards continued to draw for the same amount of time with the same passion. In another study, psychologist Edward Deci asked adults to assemble pieces of a Soma puzzle (blocks that have to be arranged in given configurations) over the course of three sessions. The group that was paid to participate in the second session (and only that session) were significantly less engaged in trying to solve the puzzle in the third session than they were in the first, whereas the unpaid group hadn't lost any of their motivation.[5] Many other studies, in a wide variety of settings with subjects of all ages, have shown that when a particular activity requires thought or creativity, extrinsic rewards almost always disincentivize the behaviour that the rewards are supposed to motivate.

In contrast to activities that are extrinsically motivated, activities that give people a sense of purpose (because they are meaningful or valuable in themselves) or that allow people to feel a sense of mastery or pride at overcoming a challenge are "intrinsically" motivating. Many studies have shown that people feel a deep sense of purpose or fulfillment when they are engaged in activities that allow them to create or discover or experience something new. According to psychologists Edward Deci and Richard Ryan, whose work inspired *Drive*, humans have an "inherent tendency to seek out novelty and challenges, to extend and exercise their capacities, to explore and to learn."

These drives help explain why students become so absorbed in their work when I guide them to figure out why a mathematical procedure works or give them a series of incrementally harder problems. Children love exploring new ideas for themselves or discovering new things, and they also love mastering new skills and demonstrating that they can surmount any challenge. I've seen this intense passion for mastery even in very young students when, for instance, I teach five- and six-year-olds to add by counting on. I tell my students that to add 4 plus 3, they just have to say "four" with their fists closed and then start counting up from 4, raising one finger for each number they say.

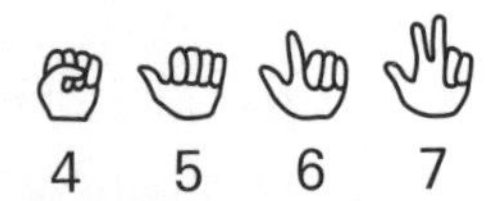

To add 4 + 3, say 4 with your fist closed. Then count up, raising one finger at a time until you have 3 fingers raised.

When children learn to add this way, they love to show off by counting on from larger and larger numbers. A drawback with this method is that students will sometimes say the first number with their thumb up (so they get an answer that is one too high). But a teacher from England once told me he had a solution for this problem. He would say the first number in the sum out loud and then pretend to throw the number to a student. The student would catch the number in their fist, with their thumb tucked under their fingers, and then repeat the number. Then they would count on. Catching the number helped them remember to start with their thumb tucked in. I've played this game with hundreds of students and I've always found it amusing to see how excited the students get just catching the numbers, even if they don't do any adding. They seem to love demonstrating that they can catch bigger and bigger numbers. And if they can count on from that number, even better.

Ben Barkley, who is principal at an American Indian school in upstate New York, recently sent me an email describing the way his grade one students (and his teachers) reacted to the game:

> *The confidence builders have already been a huge success with our teachers. The excitement and engagement was thrilling to see. People are impressed. 924 + 4 was used in 1st grade where they were catching numbers!! The kids were screaming to get into the thousands, but time ran out. One teacher was ecstatically "talking my face off" for over 5 minutes. The kids want more and had trouble staying in their seats.*

Melanie Greene, who teaches at an inner-city school on the Lower East Side in Manhattan, published a blog on the Student Achievement Partners website about the excitement her grade four students feel about math. Her school got the greatest gains on its state test math scores of all schools in New York City when it adopted JUMP in 2014. From her blog:

> *What I saw in those scores wasn't even the full picture. Something special was happening in my classroom. Each day, my students could not wait to begin math. Even my lowest-achieving students were jumping out of their seats to answer questions. I will never forget one student in particular who cried at the beginning of the school year because math was so difficult for her. She quickly got on board with JUMP Math and received a four (the highest rating) on the New York State Test that same year. Thinking of her achievement still brings tears to my eyes.*

Teachers who want their students to value math (or any subject) need to look closely at the connection between mastery and intrinsic motivation. In schools where every student is expected to master math, students get the message that math is worth learning, because their teachers are making a concerted effort to make sure that everyone does indeed learn math. This message is reinforced by the attitudes of their peers, because children love to master things, especially when they are all succeeding together. On the other hand, in schools where mastery is not a priority, students who struggle are likely to come to one of three conclusions, either consciously or subconsciously:

1. Math matters, but I can't learn it.
2. Math matters and I can learn it, but either no one can or no one wants to teach me.
3. Math doesn't matter.

None of these outcomes will help students develop an intrinsic motivation to learn math or even develop a positive relationship with their teacher. Teachers who aren't given the means to teach to mastery suffer as much as their students, because they have less satisfying relationships with students and less intrinsic motivation to teach.

Although I believe that all students (with the exception of severely learning disabled children) have roughly the same capacity for learning and engagement in math, I also recognize that some students have different short-term needs. Some students may be missing foundational knowledge or have behavioural problems or anxieties about learning. Schools need to find ways to give these students extra support and extra time to master the essential skills and concepts that they need to keep up.

But while I believe that some students need different types of support and instruction, I also believe that teachers typically "differentiate" instruction far too much. They break students into visible ability groups and have different expectations of students depending on whether, for instance, the student is a "kinesthetic" learner or a "visual" learner. They give students "low-floor, high-ceiling problems" that have "multiple entry points," because they believe that some students are only able to work on trivial problems while others will work at a very high level. They administer tests that some students aren't prepared to write and they constantly communicate to students—either overtly or in more subtle ways—that they are different. Mary Jane Moreau was able to shift the bell curve in her class so dramatically because she made all of her students feel like they could accomplish *roughly* the same things.

According to educational researcher Deborah Stipek, studies show that "students as young as first graders are well aware of the different treatment that relatively low and high performing students receive from teachers."[6] If students have an innate gift for anything, I would say it's for knowing what floor they are on in, for example, a low-floor, high-ceiling problem. If they think they have been relegated to the basement, their brains won't work nearly as well as they will if they think that everyone is enjoying life together in the penthouse. Low-floor, high-ceiling problems can sometimes be useful for occupying faster students, but they should be used with caution, so students don't know what floor they are on. As well, teachers shouldn't be satisfied to have some students permanently working on the bottom floors.

In math, I have found that it is relatively easy to keep students working on the same problems. Because I progress in manageable steps and make sure every student has the prior knowledge they need to participate in the lesson, all students can usually keep up (unless they are several grades behind and need more time on task). I differentiate my instruction by using bonus questions—as I will explain below. If my bonus questions are only small variations on the regular questions, faster students can work independently and I can usually inspire the (initially) weaker students to engage more deeply in the lesson because they see that with a little effort they can tackle the bonus questions too. I like to say that JUMP provides differentiated instruction without producing differentiated outcomes.

Teachers are sometimes reluctant to teach in a more equitable way because they are afraid that their stronger students will be shortchanged. But teachers like Moreau have shown that students can go further together. In inequitable classrooms, students who are initially stronger are held back by their classmates, who in turn progress at a much slower pace than they would in a more equitable class. As well, stronger students may be distracted by the disruptive behaviour of students who have decided they are not good at math. And when stronger students see that weaker students must struggle to learn concepts that come easily to them, they often begin to think that being good at something means not having to work. Carol Dweck's work has shown that these students are at risk academically. As school becomes more difficult, they will often give up when they encounter a challenge because they believe that they have reached the limits of their talent.

Stronger students are also often motivated to learn things for the wrong reasons. They will work to be better than their peers, to receive a higher rank or to please an adult, but not because they enjoy learning for its own sake. Because these students are only driven by extrinsic rewards that have nothing to do with the pleasures of thought, they can lose their motivation to think altogether. They will work just enough to give the illusion that they are thinking, because that is what their teachers or parents expect them to do. This doesn't happen to all stronger students, but it is a significant risk and it may explain why so many students who are good at math in elementary school gradually lose interest in the subject.

The sociologist Durkheim once observed that people rarely feel more intense excitement than when they experience the same sense of purpose or awe in a group. He called the contagious euphoria that can sweep through a crowd "collective effervescence." If we are going to teach students in groups, we might as well exploit the one advantage that comes from learning in a group. When students all conceive of an idea or master a challenge at the same time, they can be swept up in a form of collective effervescence. The excitement makes every student feel that math is intrinsically interesting and worth learning.

According to Daniel Pink, people are not only driven by a desire for mastery and purpose; they also have a drive for autonomy and like to feel that their activities are self-directed and under their control. This presents a challenge for teachers, because, as I have argued, research also shows that students need a great deal of guidance to attain mastery.

Fortunately, the pleasures of mastery seem to more than compensate for the necessity of guidance. Students enjoy being guided by a teacher, as long as they are kept in a zone where they can master the challenges. And even when teachers are guiding students, they can do many things to help students feel that they are in charge of their learning. For example, when I assess my students, I usually write on one side of the board several questions that I want every student to answer (because they can't go on otherwise). I tell them that if they complete these questions, they get a perfect score on the assessment. I then write the bonus questions on the other side of the board, and I tell students that they can do the bonus questions if they want to. Surprisingly, students almost always elect to do the bonus questions if they are able to.

Students also know that I don't give assessments or tests to rank them against their peers or motivate them to work out of fear. They know tests are for practice and that I won't give them a test if they aren't prepared to write it. I also tell them that tests help me see if I taught the material well and that if they don't do well, it may be my fault. When tests are given with the expectation that everyone will achieve mastery, they are intrinsically motivating; students see them as opportunities to show off their mastery. I once had the opportunity to teach a grade three class for five consecutive weeks. At the end of that time, I gave them a test on fractions that was well beyond their grade level and took at least thirty minutes to write. The students who had to miss the test, because they were absent that day, implored me to allow them to write it.

I am not advocating that students not be allowed to struggle. But research on motivation suggests that people will persevere longer at a difficult task when they are engaged in activities that they find intrinsically motivating. As students develop the capacity to tackle harder problems, their sense of mastery becomes a tonic that compels them to struggle more and reach higher levels. However, when students continually fail to attain mastery, they enter a vicious cycle where each failure makes their brains work less efficiently and makes them less motivated to work (even if the teacher tries to change their behaviour with extrinsic rewards or threats).

To help students feel a sense of autonomy, teachers can also set aside time for students to direct their own activities. The JUMP resources contain games and activities that allow for self-directed play. Mary Jane Moreau would sometimes ask students to select projects that they could work on at home; for example, after teaching a lesson on probability, she asked the students to design their own game. But she would never send home homework on topics that students needed to know. She didn't want to create inequalities in her class by asking parents to supervise homework or teach concepts or skills the students would need the next day. She knew that some parents might not have the time or expertise required to help their kids with math. (If teachers do want to assign homework, I recommend that they send home material the students already know but need to practise.)

The Power of Incremental Variation

When I first started volunteering in inner-city classes, it occurred to me (perhaps because I am a playwright) to think of the students as an audience. As Durkheim pointed out, people never get more excited or feel things more intensely than when they are in a group, experiencing the same thoughts and emotions or following the same story. In all of the discussions about educational reform that I have heard, I have never heard anyone talk about this "audience effect." This may be because people find it hard to imagine an entire class getting excited about math or succeeding at the same level.

When there are visible academic hierarchies in a group, learners who feel inferior will often disengage, so there will be little collective excitement in the group.

But this puts many learners at a serious disadvantage, because our brains work better when we are excited about what we are learning.

Most teachers find it hard to keep their students in a zone where they are all working to meet the same challenges or understand the same concepts. That's because students have different levels of background knowledge and work at different rates. Even when I am teaching in easy steps, some of my students will invariably need a little extra time to practise a skill or consolidate an idea. In these cases I create extra bonus questions, so that the students who have mastered the step can work independently while I focus on the ones who need my attention.

When teachers want to assign extra work to their stronger students, they usually use questions that they find in textbooks or on math websites. Because teachers are so busy, they don't always have time to analyze these questions carefully to see if they contain any new vocabulary, skills or concepts that students haven't learned. When the questions require prior knowledge that hasn't been taught, the teacher usually has to spend time helping stronger students and will have less time to devote to students who really need the help.

When I create bonus questions, I'm careful not to vary too many elements of the problem or to introduce any new skills or concepts. My goal in creating these questions is to challenge the faster students but not to stump them. The bonus questions should allow me to spend more time with students who really need my help, while stronger students work independently. A well-designed sequence of bonus questions can also benefit students who are initially weaker in another way: when these students see that the bonus questions are within their grasp, they start to focus more and work harder so that they can get their bonus questions too.

Here is an example of a continuum of bonus questions that would keep faster students productively occupied and also provide a path to deeper understanding that less advanced students can follow: In a fraction, the denominator of the fraction (or the number on the bottom) tells you how many pieces are in one whole. The numerator tells you how many pieces you are interested in or how many you have selected. To add a pair of fractions, you add the numerators because you want to know how many pieces you have selected altogether in both fractions. But you don't add the denominators. If you eat one sixth of a pizza and then eat another sixth, you will have eaten two sixths of a pizza. Notice that when you add $\frac{1}{6}$ plus $\frac{1}{6}$ and get $\frac{2}{6}$, the denominator doesn't change, because the size of each piece of pizza doesn't change. Similarly, when you subtract a pair of fractions, you subtract the numerators (because you are taking away pieces), but you keep the same denominator.

When students have learned how to add and subtract simple pairs of fractions, such as $\frac{1}{3} + \frac{1}{3}$ and $\frac{5}{8} - \frac{1}{8}$, I will begin to write a sequence of bonus questions on the board for students who need extra work. I might start by making the denominators bigger:

$$\frac{1}{325} + \frac{3}{325}$$

(Surprisingly, young students think this is a harder question than $\frac{1}{4} + \frac{1}{4}$, even though the denominator doesn't play any role in the addition. When I increase the size of the denominator to be a number in the thousands or ten thousands, even grade five students get excited.)

I might also ask students to add three fractions, or I might combine addition and subtraction in the question.

$$\frac{1}{7} + \frac{1}{7} + \frac{1}{7} \qquad \frac{3}{10} + \frac{4}{10} - \frac{2}{10}$$

I might make a mistake and ask students to correct my mistake.

$$\frac{2}{11} + \frac{5}{11} = \frac{7}{22}$$

I might create an algebraic puzzle where the student has to fill in a missing number.

$$\frac{\square}{13} + \frac{5}{13} = \frac{9}{13} \qquad \frac{12}{17} - \frac{\square}{17} = \frac{6}{17}$$

I might even push students to think outside the box a little—without introducing any new concepts. When I ask students to simplify the following expression, they often protest and say that they can't do anything with the expression because I haven't taught them to add fractions with different denominators.

$$\frac{1}{3}+\frac{1}{4}+\frac{1}{5}+\frac{2}{3}+\frac{3}{4}+\frac{4}{5}$$

I tell students not to give up, because they have all the skills they need to solve the problem. Eventually they see that they can find the answer by changing the order of the terms and adding terms with the same denominator.

$$\frac{1}{3}+\frac{1}{4}+\frac{1}{5}+\frac{2}{3}+\frac{3}{4}+\frac{4}{5}$$

$$=\frac{1}{3}+\frac{2}{3}+\frac{1}{4}+\frac{3}{4}+\frac{1}{5}+\frac{4}{5}$$

$$= 1 + 1 + 1$$

$$= 3$$

Students of all ages like meeting incrementally harder series of challenges (as in a video game) and they like being able to show off in front of their peers. That is why it's important to make the first questions in a continuum of bonus questions easy enough to seduce the weakest students into following the sequence of challenges. If I find the right entry point for the weakest student, I know the stronger students will eventually benefit when the whole class speeds up and starts enjoying math together. Because students' brains work more efficiently when they are excited, this speed-up can happen *very* quickly—often in one lesson.

It isn't always easy for teachers to create sequences of questions where only one or two features of the questions change at a time. Mary Jane Moreau loved teaching math and was recognized as an excellent teacher before she started using JUMP. But after reading the JUMP teachers' guides, she said she realized that many of the concepts she had previously taught in one step actually involved three or four steps or required skills or knowledge that she didn't normally assess or teach. The JUMP writers and I have spent a great deal of time learning how to teach in small steps, but even with all of our experience, we still sometimes vary too many elements of a problem at the same time without noticing.

In an early edition of the JUMP student books, we created several diagrams that were supposed to help students learn to recognize pairs of parallel lines. In every diagram, the two lines were the same length and one line was situated directly above the other, as shown below.

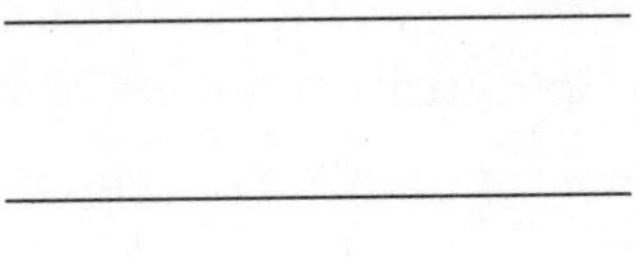

However, on our quizzes, we asked students to say whether the pairs of lines in Figures A and B were parallel.

A

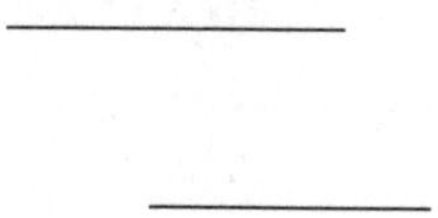

B

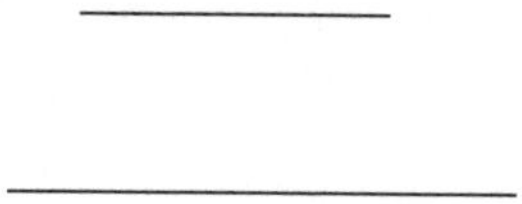

Some students thought that neither of these pairs of lines were parallel. They surmised, based on the examples in the book, that parallel lines had to be the same length and had to be aligned so that neither line extended beyond the other in either direction.

This example shows why experts don't always make the best teachers. People who have spent a great deal of time learning how to recognize different instantiations of a concept sometimes have trouble appreciating how different those instantiations might look to a novice. The JUMP writers and I thought that the examples of parallel lines in the student book and in the quizzes were similar, because in every example the lines looked (to us) like they would not meet if they were extended indefinitely. But by varying too many things at once, including the relative length and orientation of the lines, we inadvertently created a set of examples that looked, in the eyes of a novice, very different from one another. Parents, too, when they think their children are being wilfully obtuse or ignoring instructions that they believe are obvious, should remember how hard it is for novices to follow too many conceptual variations.

The research on learning styles and motivation that we looked at in this chapter is a small part of a large body of research on human behaviour (including research on anxiety, empathy and executive function) that has important implications for equity in education. For example, a recent study by psychologist Sian Beilock and her colleagues suggests that young students will internalize the math anxiety of a teacher who is of the same gender more than they will internalize the anxiety of a teacher of the opposite gender.[7] Because the majority of elementary school teachers are female and because a high proportion of elementary school teachers are math phobic (this is well established), girls may experience the negative effects of teachers' anxieties about math more than boys. Young girls generally do as well as or outperform boys in math now, but they have more negative perceptions of their level of achievement than boys and are less likely to specialize in STEM subjects when they are older. The new research on anxiety may help explain these gender differences and empower us to find ways to give boys and girls the same opportunities to pursue careers in math or fields that require math.

In Moreau's class, there was no significant difference between boys and girls when it came to achievement in or enthusiasm for math. This is not surprising, as the research on learning I have discussed suggests that we could—if we were serious about equality—eliminate the opportunity gap in math with very little effort.

Humans have a long history of seeing stark differences between people even when those differences are illusory or superficial. The American slave trade was grounded in the belief that African Americans had the intellectual capacity of children, and patriarchal societies traditionally believe women aren't suited to higher-level education or work outside the home. Most varieties of inequality take one of two forms: a group of people who enjoy more opportunities than another group (in the previous example: white men) will decide, arbitrarily, that the other group either doesn't desire the same opportunities or isn't capable of benefiting from the same opportunities. Our attitudes about mathematical abilities conflate both of these forms of ignorance.

The idea that many people are not predisposed to *like* math is as unscientific and damaging as the idea that many people are not capable of *learning* math.

Fortunately, in the past two decades, the science of motivation has converged with the science of learning. We now know that learners love mastery and that—thanks to a lucky outcome of evolution—they also learn best through mastery. In the next chapter I will argue that creativity, like higher-order thinking and problem solving, can also be developed with the right approach.

In the early 1980s, when I was first learning to write plays, I would always carry a notebook so I could write down fragments of conversations that I overheard on buses and trains or in other public places. Once, on a street in Manhattan, I saw a couple who were engaged in a very heated (and very public) argument about their relationship. As I was walking past the pair, the woman, who was seething with frustration, shouted at her partner: "You just don't know what you don't want!" I immediately wrote the line in my notebook (and eventually used it in one of my plays) because I thought it was both funny and illuminating. The tortured syntax of the sentence, with the double negative, perfectly captured the woman's frustration. But it also suggested a deeper truth about human experience that hadn't occurred to me until then. We all spend a great deal of time trying to figure out what we want in life, but we probably spend even more time learning, by a somewhat random and often painful process of trial and error, what we don't want.

Many of the best lines or scenarios in my plays are based on things I've seen or heard that I couldn't have imagined if I hadn't witnessed them. When I was young, I thought (as many young people do) that artists and scientists generate all of their ideas entirely from their own imaginations and that those ideas arrive fully formed. But, in the process of becoming a playwright and mathematician, I learned that creativity more often involves selecting and organizing material that is presented, somewhat haphazardly, by your imagination or by the world. As the philosopher Friedrich Nietzsche explains:

> *Artists have a vested interest in our believing in the flash of revelation, the so-called inspiration . . . [shining] down from heavens as a ray of grace. In reality, the imagination of a good artist or thinker produces continually good, mediocre, and bad things, but his judgment, trained and sharpened to a fine point, rejects, selects, connects. . . . All great artists and thinkers [are] great workers, indefatigable not only in inventing, but also in rejecting, sifting, transforming, ordering.*[1]

In the process of becoming a writer and mathematician, I also learned that structure is not an impediment to creativity; it often enables creativity. Mathematicians have found ways to bend the seemingly rigid rules and definitions of mathematics into startlingly original configurations, and writers have used the restrictions of their medium (for example, the strict metres or rhyme schemes of classical poems and plays) to spark their imaginations and to help them shape their thoughts.

The haiku is a particularly difficult poetic form, because the poet's choices are tightly constrained by the number of syllables in each line. But one of my favourite haikus was written by an elementary school student who used those constraints to create a very funny self-referential poem that was also an act of protest. When the boy's teacher told him he had to write a haiku for his creative writing assignment, this is what he came up with:

> Five syllables here
> Seven more syllables there
> Are you happy now??

In this chapter, I will look more deeply at the techniques and habits of mind that artists and scientists have drawn on to sift, order and transform their experiences into original scientific theories or works of art. I will also discuss the wider role that creativity and curiosity play in success in every sphere of life.

Mathematics and Creativity

Just as there is an emerging science of expertise, there is also an emerging science of creativity. As Nietzsche foresaw, this research suggests that creative people don't wait passively for divine sparks of inspiration to strike them but instead are good at generating (or collecting) a large volume of ideas and using their expertise to select the ones that are likely to be fruitful. They often use trial and error to find solutions to problems, and they are extremely persistent. Beethoven would sometimes run through as many as sixty or seventy different drafts of a phrase before settling on the final one. "I make many changes, and reject and try again, until I am satisfied," the composer told a friend. "Only then do I begin the working-out in breadth, length, height and depth in my head."[2]

Highly creative people tend to have an extremely wide range of interests and hobbies. One study found that most Nobel Prize winners in physics and chemistry were also accomplished writers, musicians or artists. Having an expertise in many domains can help creative people think outside the box and see analogies and connections between things that may not seem on the surface to be related. For example, according to psychologist Dean Simonton, "Galileo was probably able to identify lunar mountains because of his training in the visual arts, particularly in the use of chiaroscuro to depict light and shadow."[3]

In his book *Originals*, Adam Grant describes a new body of research that suggests that highly creative people are often not the best judges of the merits of their own work and that they will frequently value their less important work over their most important. However, they *are* good at judging the value of the work of their peers (while experts from other fields are less capable).[4] One way creative people can increase their odds of success is to engage in constant exploration, while drawing on the expertise of peers to help them decide where to focus their energy.

According to Grant, highly creative people tend to be inspired by stories of invention and adventure that they encountered in their youth. One large-scale study has provided intriguing evidence for the impact that stories can have on creativity: in the United States, a significant increase in the number of children's stories that emphasize original achievements (from 1800 to 1850) was followed by a steep increase in the number of patents granted (from 1850 to 1890). Parents and teachers can foster creativity by introducing children to these kinds of stories when they are young. My interest in both the arts and the sciences was definitely sparked by stories that I read about artists and scientists when I was growing up.

Creativity and curiosity are closely linked. Creative people are driven to seek out new experiences and new knowledge and will work relentlessly on puzzles and problems that they feel compelled to solve. Leonardo da Vinci was not only a universally talented creative genius; he was also, as art historian Kenneth Clark said, "undoubtedly the most curious man who ever lived." In a passage from his notebooks, Da Vinci writes:

> *I roamed the countryside searching for answers to things I did not understand. Why shells existed on the tops of mountains along with the imprints of coral and plants and seaweed usually found in the sea. Why the thunder lasts a longer time than that which causes it and why immediately on its creation the lightning becomes visible to the eye while thunder requires time to travel. How the various circles of water form around the spot which has been struck by a stone, and why a bird sustains itself in the air. These questions and other strange phenomena engage my thought throughout my life.*[5]

Drawing on over forty years of research on curiosity, psychologist Todd Kashdan and his colleagues at the Center for the Advancement of Well-Being at George Mason University have identified a number of traits that highly inquisitive people seem to have in common. These traits include a willingness to accept and harness the anxiety associated with novelty; a willingness to take physical, social and financial risks to acquire new experience; an interest in observing others to learn what they are thinking and doing; a drive to fill in gaps in knowledge; and a capacity to be in a state of wonder and to feel pleasure at being in that state.[6]

It's easy to see that scientists and artists are more likely to be successful in their fields if they are deeply inquisitive. Psychologists have found that curiosity can also enhance the quality of our lives in many ways. According to Kashdan,

> *Psychologists have compiled a large body of research on the many benefits of curiosity. It enhances intelligence. In one study, highly curious children aged three to 11 improved their intelligence test scores by 12 points more than their least-curious counterparts did. It increases perseverance or grit. Merely describing a day when you felt curious has been shown to boost mental and physical energy by 20% more than recounting a time of profound happiness. And curiosity propels us toward deeper engagement, superior performance and more meaningful goals. Psychology students who felt more curious than others during their first class enjoyed lectures more, got higher final grades and subsequently enrolled in more courses in the discipline.*[7]

In "The Business Case for Curiosity," psychologist Francesca Gino presents evidence that curiosity produces a wide range of benefits for organizations, leaders and employees. For example, in a state of curiosity, we are less susceptible to confirmation biases (looking for information that confirms our beliefs rather than evidence suggesting we are wrong) and to stereotyping people (making generalizations such as that women or people of colour don't make good leaders). Instead, curiosity leads us to generate alternatives.

According to Gino, higher levels of curiosity among employees can lead to more innovation on the job, to reduced group conflict (curiosity encourages members of a group to put themselves in one another's shoes and take an interest in one another's ideas), and to more open communication and better team performance (groups whose curiosity is heightened share information more openly and listen more carefully). Psychologists have found that the effects of curiosity also extend to higher levels of management. For example, the executive search firm Egon Zehnder found that curiosity is the best predictor of strength in all of the leadership competencies that the firm measures; these competencies include strategic orientation, collaboration and influence, team leadership, change leadership and market understanding.[8]

Many people believe, based on their experience of learning math at school, that math is a rigid and sterile subject that stifles curiosity and leaves no room for creativity. But progress in mathematics has actually been driven by remarkable flights of imagination. Math can be an ideal tool for nurturing curiosity in learners of all ages.

Humans appear to have evolved to enjoy solving puzzles and posing problems. Sixty years ago, Henry E. Dudeney, who was a great creator and compiler of puzzles, observed that an innate "curious propensity for propounding puzzles" shows itself in different forms across cultures and historical periods.

Around the world, millions of people spend countless hours creating and solving puzzles ranging from crosswords to Sudoku puzzles, to board and card games. And this "curious propensity" for posing and solving problems is not unique to our species. As psychologist Frank Dumont points out, monkeys will spend time solving puzzles (involving mazes for example) for which there is no cued reward "just for the fun of it."[9]

Part of the pleasure that comes from solving a puzzle or problem undoubtedly has to do with the sense of mastery you feel at conquering it. But to solve a puzzle, you must synthesize or combine the information in the clues to create new information or generate new ideas. A wide range of research, from psychology to evolutionary biology, suggests that this process itself is also part of the reward of solving the puzzle.[10] Humans and other primates appear to be compelled, by very primitive drives, to seek new information and reduce uncertainty. For example, research has shown that young children will structure their play to gain new information and make causal connections. And when monkeys are offered two different ways of activating the same reward (where in one case they will receive information about the size of the coming reward and in the other they won't) they consistently make the choice that gives them the most information.[11]

Most of the puzzles that we encounter in daily life are not well suited for fostering our natural curiosity. They aren't designed to draw our attention to the structures that we need to see to solve the problem. And many puzzles require too much domain-specific knowledge or expertise for the average person to solve them. I, for one, am too intimidated to try the cryptic crosswords that appear in papers like the *Globe and Mail* or the *New York Times*. People often lose the inquisitive spirit they had as children because they are required, at school and in life, to engage with too many problems that they aren't equipped to solve. But this needn't be the case in mathematics. In math it is easy to create well-scaffolded sequences of challenges that allow people to experience the pleasure that comes from making discoveries. People can enhance their curiosity and their capacity to persevere by solving problems.

Advanced Sudoku puzzles are challenging, but it's not hard to design a version of Sudoku that a novice can enjoy and that provides a path for them to learn to play the full game. A Sudoku puzzle consists of 9 three-by-three grids that are partially filled in with one-digit numbers (excluding zero), as shown on page 201. The goal in the puzzle is to fill in the missing numbers so that every one of the three-by-three grids, as well as every row and every column, contains each of the numbers from 1 to 9.

Here's a typical Sudoku puzzle:

5	3			7				
6			1	9	5			
	9	8					6	
8				6				3
4			8		3			1
7				2				6
	6					2	8	
			4	1	9			5
				8			7	9

One way to make the puzzle easier, without omitting any of its essential features, is to use two-by-two grids (as shown below) and to require that each grid, row and column contain only the numbers from 1 to 4.

1			4
4	3	2	
	4	1	3
3		4	

	4	2	
2			4
4			1
	3	4	

	4		3
1			
	2		

		4	
3			
	2		2

Some 2 by 2 puzzles.

To help a novice understand the rules and develop the basic strategies they need to solve a Sudoku puzzle, a teacher (or parent) can also design a series of exercises in which the numbers that are filled in draw the learner's attention to critical features of the puzzle.

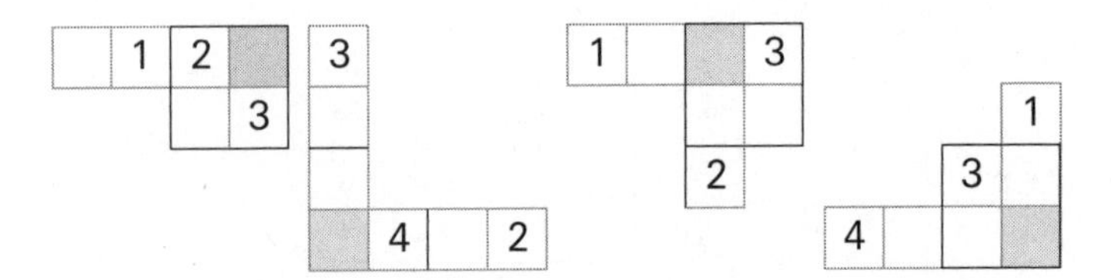

Here are some examples of warm-up exercises that can help students learn to play Sudoku.

Of course, this approach to teaching doesn't only work for isolated puzzles. All mathematical concepts can be taught through structured inquiry. And effective strategies for generating creative solutions to problems in any domain, including the use of analogy and abstraction, can be learned through mathematics.

The Power of Analogy

One of the most powerful methods for generating new ideas (rarely discussed in popular literature about genius and talent) is the method of drawing analogies. In physics and mathematics, many important conceptual advances in the past two hundred years were inspired by analogies.

If you have ever sprinkled iron filings around a magnet, you have seen that the filings mysteriously line up along curved arcs that join one end of the magnet to the other. The physicist Michael Faraday, whose experiments laid the foundations for the modern theory of electromagnetism, called these arcs "lines of force." He believed that space is permeated with these lines and that areas of space where the electromagnetic force is weaker contain fewer lines.

Many physicists of his time were dubious about this idea. They preferred to think of space as an empty vacuum and believed that electrically charged objects could affect each other at a distance without any intervening physical medium. They claimed that Faraday wasn't qualified to develop a theory of electromagnetism because he didn't know enough mathematics. But one mathematically inclined physicist, James Clerk Maxwell, was able to give Faraday's ideas a proper theoretical foundation. He did this by finding an analogy that allowed him to use ideas from another branch of physics to make Faraday's theory work. As Maxwell's biographers, Lewis Campbell and William Garnett, explain: "Taking an illustration from the flow of water in a river, Maxwell pointed out that the stream lines or paths along which particles of water flow, are analogous to lines of electric force, the velocity of the water being analogous to the intensity of the force."[12]

By reimagining Faraday's lines of force as a collection of infinitesimal tubes that direct the flow of force, like water in a river, Maxwell was able to use the mathematics that had been developed to describe the behaviour of fluids to derive all of the key equations that govern the behaviour of magnets and electric currents. He was also able to determine the speed of light by showing that light is made of electromagnetic waves that undulate through space (another metaphor). It's hard to overstate the importance of these analogies. As physicist Richard Feynman put it: "From a long view of the history of the world—seen from, say, ten thousand years from now—there can be little doubt that the most significant event of the 19th century will be judged as Maxwell's discovery of the laws of electromagnetism."[13]

The general theory of relativity was also born from an analogy. One day when Einstein was struggling to understand how gravity could be accounted for in his special theory of relativity (which explains how the world works in the absence of gravity), he noticed some workers repairing the roof on a building across the street. He was momentarily distracted by the thought that one of the workers might fall, but then he had what he described as the happiest thought of his life.

Einstein wondered if a person who was plummeting toward the surface of the earth, and who also happened to be enclosed in a box without windows, would be able to tell that they were falling under the influence of gravity. He realized that this person would experience the same feeling of weightlessness as someone who was confined to a stationary box in a region of space with no gravitational field. Neither observer could tell if they were standing still in outer space or were in "free fall," accelerating under the force of gravity. Similarly, a person who felt their feet pressing against the floor of a box wouldn't know if their box was accelerating upwards in a region of space with no gravitational field or if their box was standing still in a gravitational field. This comparison between boxes in different regions of space helped Einstein see how gravitation could be included in the theory of relativity.

Einstein used a different analogy (where he imagined a beam of light made up of tiny packets of energy) to show that Maxwell's wave picture of light was incomplete. These packets of light were eventually named photons, and Einstein's analogy provided a basis for the theory of quantum mechanics and the strange idea that light can behave both like a particle and like a wave.

A mathematical breakthrough that underlies much of modern physics is also based on an analogy. In mathematics, the numbers that humans discovered (primarily) through our interactions with the world are called the "real" numbers. The whole numbers, the fractions and the integers are all examples of real numbers. In the real number system, it doesn't make sense to take the square root of a negative number, because when you multiply a real number by itself (regardless of whether its sign is positive or negative), the result is always a positive number. But during the Renaissance, mathematicians found that if they allowed themselves to take square roots of negative numbers, they could solve equations that were too hard to solve by any other methods. They called expressions with negative square roots "imaginary" or "complex" numbers, because, even though they didn't have tangible equivalents, they could still be manipulated like real numbers.

At this point you might wonder why mathematicians would choose to call expressions that have no meaning in the real number system "numbers." There's a deep reason why this is the correct name for these strange entities. In developing the idea of complex numbers, mathematicians created a surprising and powerful analogy. And this analogy changed our understanding of what it means for something to be a number.

To see why imaginary numbers deserve to be called numbers, let us consider some basic properties that all numbers have in common. In chapter two, we encountered a set of axioms that were discovered in ancient Greece from which much of geometry can be derived. In the 1800s, mathematicians began to search for a set of axioms that could provide a similar foundation for number theory. They soon discovered a set of simple properties from which all other properties of numbers could be derived. One of these properties is called the "commutative property" of addition. If you add a pair of real numbers, you get the same answer regardless of the order in which the numbers are written: 3 plus 4 is the same as 4 plus 3. Numbers also commute when you multiply them. The properties that define the real numbers are all that simple.

In the mid-1800s, Irish mathematician William Hamilton suggested a way of looking at complex numbers that makes their resemblance to real numbers easy to see. He abandoned the use of negative square roots and said that every complex number should be represented by an ordered pair of numbers written in the way that coordinates on a grid are normally written. In Hamilton's notation, (1, 5) and (2, 3) are examples of complex numbers. Hamilton also defined a way of adding and multiplying these numbers. To add a pair of complex numbers, you simply add the numbers in the first position to each other and the numbers in the second position to each other. So, for example:

$$\begin{aligned} &(1, 5) + (2, 3) \\ &= (1 + 2, 5 + 3) \\ &= (3, 8) \end{aligned}$$

The rule for multiplying complex numbers isn't as simple as the rule for addition: it mixes up the numbers in the first and second positions in a somewhat complicated way that is shown in the appendix by this rule:

$$\begin{aligned} &(1, 5) \times (2, 3) \\ &= (-13, 13) \end{aligned}$$

It's easy to see that complex numbers commute under addition, just as the real numbers do.

$$(1, 5) + (2, 3)$$

$$= (1 + 2, 5 + 3)$$

$$= (2 + 1, 3 + 5)$$

$$= (2, 3) + (1, 5)$$

In the expression above, the first equal sign holds by the definition of complex addition, the second holds because ordinary numbers commute under addition, and the third holds by the definition of complex addition. In the appendix, you can see why complex numbers also commute under multiplication.

It's easy to prove that complex numbers satisfy all of the same basic properties that real numbers do. And since complex numbers are algebraically indistinguishable from real numbers, we can use them for the same purposes. It's possible, for instance, to do calculus with complex numbers. But when we use complex numbers in calculus, something magical happens. Many calculations and proofs that are extraordinarily hard to do with ordinary numbers turn out to be trivial with complex numbers.

The numbers we experience in the world, the real numbers, are a subset of the complex numbers. Or more precisely, the real numbers are just the complex numbers that have a zero in the second position. For example, the numbers 3 and 4 are simply the numbers (3, 0) and (4, 0) in a different notation. When we add (3, 0) and (4, 0) we get (7, 0), which is equivalent to 7. Normally the numbers in the first and second positions in a pair of complex numbers are mixed up when the numbers are multiplied, but this does not happen when there is a zero in the second position of both complex numbers. When we use the rule for complex multiplication to multiply (3, 0) by (4, 0) we get (12, 0), which is equivalent to 12. The real numbers are effectively a one-dimensional slice or subset of a two-dimensional number system.

Complex numbers make many of the calculations in modern physics possible. But that is not what makes them truly magical. Even though we never encounter complex numbers in our lives, the operations of the universe are actually governed by complex numbers rather than real numbers. For instance, if you try to predict how a stream of electrons will behave in a magnetic field, your predictions will only be accurate if you use complex numbers to calculate the probabilities.

I always feel a sense of awe when I think that the numbers that underlie the design of the universe weren't discovered through experience or experimentation, but by pure thought. Using analogical reasoning, mathematicians extended the concept of number to a class of abstract entities that we never experience directly but that underlie all of reality. To me this represents one of the most remarkable examples of human ingenuity.

To further unlock the potential of human ingenuity, I believe we need to demystify the notion of intellectual talent in two ways. To start with, let's abandon the idea that the average person is incapable of understanding the deepest and most beautiful ideas in math and science. Undergraduate physics students now have a deeper understanding of some aspects of the theory of relativity than Einstein did. And the research I've discussed in this book suggests that just about anyone could learn the math that's needed to do undergraduate physics.

Einstein's genius didn't lie in having ideas that are inaccessible to other minds. He was a genius because he *discovered* those ideas. But even if we are clear about this, we still need to go further to demystify the idea of genius. Let's also abandon the idea that the average person isn't capable of making interesting or useful discoveries. Of course, it would be unreasonable to expect everyone to make the kind of world-changing discoveries we have explored in this chapter. But research in cognitive science suggests that even the highest forms of reasoning, including the use of analogies, can be learned through training and practice.

Unfortunately, the research also suggests that, unless they receive training, people aren't *naturally* very good at seeing or using analogies. That is why geniuses like Einstein are so rare. In 1980, psychologists Mary Gick and Keith Holyoak performed a classic experiment that showed that people will often overlook a solution to a problem even when an analogy that contains the solution is right under their noses.[14]

In the experiment, the researchers asked participants to try to solve the following problem, which is based on a real medical scenario: A patient has a tumour that can't be operated on. Fortunately doctors have a ray that can destroy the tumour. Unfortunately the ray also destroys healthy tissue, unless it is administered in doses that are too low to destroy the tumour.

Before they were given the problem, some of the participants read this story:

A small country was ruled from a strong fortress by a dictator. The fortress was situated in the middle of the country, surrounded by farms and villages. Many roads led to the fortress through the countryside. A rebel general vowed to capture the fortress. The general knew that an attack by his entire army would capture the fortress. He gathered his army at the head of one of the roads, ready to launch a full-scale direct attack. However, the general then learned that the dictator had planted mines on each of the roads. The mines were set so that small bodies of men could pass over them safely, since the dictator needed to move his troops and workers to and from the fortress. However, any large force would detonate the mines. Not only would this blow up the road, but it would also destroy many neighbouring villages. It therefore seemed impossible to capture the fortress. However, the general devised a simple plan. He divided his army into small groups and dispatched each group to the head of a different road. When all was ready he gave the signal and each group marched down a different road. Each group continued down its road to the fortress so that the entire army arrived together at the fortress at the same time. In this way, the general captured the fortress and overthrew the dictator.

Even though the story contains the solution to the problem—expressed as an analogy—only 30 percent of participants who read the story were able to solve the problem. However, when participants were given a hint suggesting that the story could help them solve the problem (but making no explicit reference to the analogy), the success rate rose to over 90 percent. Almost all of the participants who were given the hint saw that doctors could destroy the tumour by irradiating it with small doses of radiation from many directions at the same time. This result (where students who are guided to make comparisons dramatically outperform students who aren't) has been replicated in multiple studies in many domains.

According to Dedre Gentner, who was one of the pioneers in the study of analogical reasoning, an analogy is a mapping of knowledge from one domain (which psychologists call the "source") into another domain (called the "target"). In an analogy, objects that play similar roles in the two domains are mapped onto each other. The matched objects don't have to resemble each other, as long as they perform similar functions in their respective domains.[15]

Analogies are powerful tools for solving problems, because they reveal structural relationships between domains that are independent of the objects in which those relationships are embodied. Scientists often use analogies to apply knowledge from one field to another, even when the objects of study in the two fields seem to have very little in common. Analogies are also frequently used by teachers to help students understand new concepts.

As Gentner explains, a teacher might help students understand the atom by presenting an analogy between the solar system and the atom.

> *In this example, the solar system represents a domain that is already familiar to students (the source), and the atom represents the domain that students are learning about (the target). Appreciating and learning from this analogy requires the student to look past surface-level differences*

> *between the source and target and instead notice the underlying, shared relational structure between domains—in this case, the fact that the planets orbit the sun in an analogous fashion as the electrons orbit the atom's nucleus.*[16]

Both children and adults have trouble using analogies to solve problems, because they often find it hard to see the shared structure between two domains or to match objects in the source with corresponding objects in the target, especially if they are misled by superficial features of the objects in the two domains. For example, young children tend to over-focus on objects that look similar, regardless of what role they play in the source and target. In one experiment, four-year-olds were given two cards that each showed one large and one small object. They were then given a card that showed the two large objects from the original cards and a card that showed a large and a small object that were both new. When asked which card matched the two cards in the original set, the children almost always picked the card with the two large objects rather than the card that showed the relation larger to smaller.

In the past two decades, psychologists have discovered a variety of effective methods for helping people become better at seeing analogies. For example, studies have shown that learners who are asked to describe the similarities between two analogous contexts before being presented with a problem do significantly better on knowledge transfer tasks. In one study, Gentner and her colleagues found that "business school students who compared two negotiation scenarios were over twice as likely to transfer the negotiation strategy to an analogous test negotiation as were those who studied the same two scenarios separately."

This result has been replicated in a wide variety of domains. Other studies have shown that analogical reasoning is enhanced by visual cues. Effective approaches include displaying representations of the source and target simultaneously rather than sequentially, highlighting corresponding elements of the source and target, and using gestures.

> *Richland and McDonough (2010) provided undergraduates with examples of permutation and combination problems that incorporated visual cueing, such as gesturing back and forth between problems and allowing the examples to remain in full view, versus comparisons that did not incorporate visual cueing. Students who studied the problems with visual cueing were more likely to succeed on difficult transfer problems.*[17]

Studies have also shown that students derive the greatest benefit from comparison tasks that are well scaffolded and in which the amount of variation between the content is limited. For example, when students are asked to compare correct and incorrect solutions of an algebraic problem, they learn most efficiently when the examples are similar except for a single, key difference. In "Analogical Reasoning in the Classroom: Insights from Cognitive Science," psychologist Michael Vendetti and his colleagues present a variety of additional strategies that teachers can use to help learners become better at seeing and using analogies.[18]

Researchers have shown that students can be trained to significantly improve their performance on a wide variety of analogy problems, including the types that appear on high-stakes tests of achievement (like the SAT exams that are used in US college admissions) or intelligence tests. That learners who are guided to make comparisons show such dramatic improvements in their ability to solve problems suggests that analogical reasoning is not inherently hard, but depends more on a learner's background knowledge and habits of mind.

Because of my literary training, I habitually look for similarities or analogies, even in domains that may appear to be different or unconnected. For example, a comparison between two equations from different branches of math that looked very similar—one in a field that I was studying as a doctoral student and one in a book that I randomly picked off a library bookshelf—led me to my first discovery in mathematics. I also use a method of asking questions that I learned from the philosopher Wittgenstein that involves asking if particular concepts are necessary or the result of historical or cultural accidents. It helps me see the structures, relationships and presuppositions that are implicit in ideas. I believe that mathematicians and scientists can benefit from training in the arts (and vice versa) because people who can see analogies in a variety of domains are likely to have a greater sensitivity to invisible structural relations and hidden similarities.

The Power of Abstraction

Most analogies in mathematics involve some form of abstraction. In chapter two, I suggested that much of the progress in mathematics in the past two hundred years has been achieved because mathematicians learned to see familiar mathematical entities, such as numbers and shapes, more and more abstractly. One of the most unlikely advances in mathematics was sparked by a question that appeared in a letter to a British literary magazine in 1854. The author of the letter, identified only by their initials as F. G., asked if there is a minimum number of colours that could be used to colour any map, regardless of how many countries are in the map or how they are arranged, so that two countries that share a border are never given the same colour. Although the question sounds rather trivial, it turned out to be extraordinarily hard to answer. Most mathematicians thought that four colours were sufficient to colour any map, but, like Euclid's fifth axiom, this conjecture inspired many false proofs. In 1976 two mathematicians, Kenneth Appel and Wolfgang Haken, finally proved that this conjecture was true. Their paper was over a thousand pages long, and parts of the proof were so complicated that they could only be checked by computer.

Not long after F. G.'s question appeared in print, mathematicians saw that a structure that was already known in mathematics (but that hadn't proved very useful until then) could be used to represent a map. This form of representation, which is called a "graph," is more abstract than the map because it effectively leaves out any features of the map that aren't relevant to the solution of the colouring problem – for example, the size of a particular region or the contour of a border.

In pure mathematics, a graph is less complicated than the graphs of economic performance, for example, that you may be familiar with from the news. The graphs that economists use to represent supply and demand have horizontal and vertical axes and lines that represent various quantities, whereas a graph in pure mathematics is simply a collection of dots and lines that join the dots. The image on page 219 on the right-hand side is a graph. The dots are called "vertices" and the lines between pairs of vertices are called "edges." In this case, the vertices of the graph are labelled with colours (R for red, G for green, and so on) because the graph is meant to represent the map on the left-hand side of the picture; normally the vertices of a graph are not labelled.

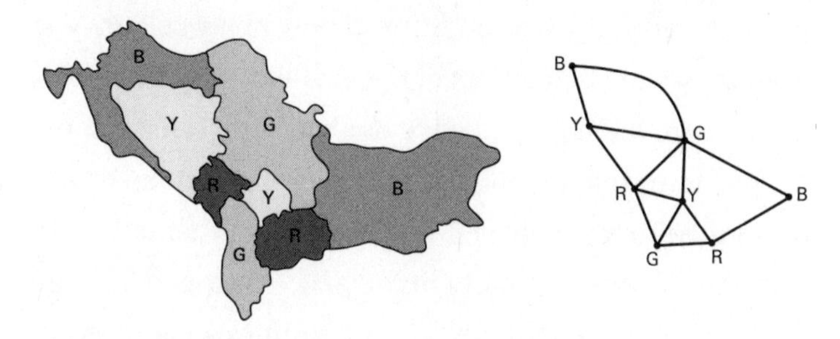

If you would like to get a sense of what it feels like to discover a mathematical analogy, you might try to figure out which elements in the map (the source) are represented by the vertices in the graph (the target) and what relationships in the map are represented by the edges in the graph. Why not try it out before you read on.

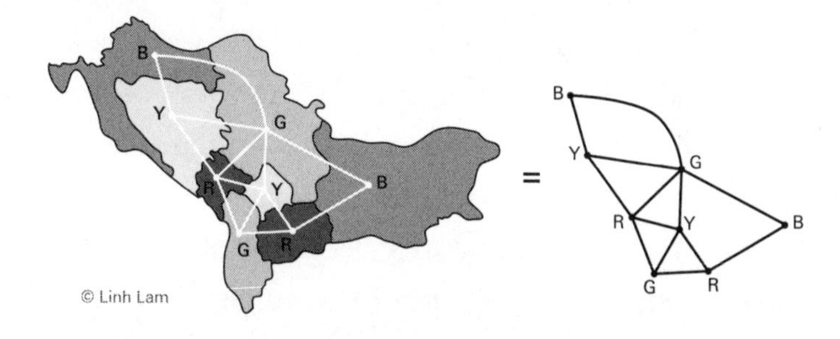

When you compared the two images, I hope you saw that each vertex in the graph represents a country on the map. Two vertices are joined by an edge if and only if the two countries represented by the vertices share a border. Above, the graph is superimposed on the map to make the connection between the source and the target clear. On graphs that represent maps, the colouring problem reduces to the question of how many colours you need to colour the vertices of the graph so that two vertices that are joined by an edge don't get the same colour.

The four-colour problem is a particularly good example of the unreasonable effectiveness of mathematics. To solve the problem, mathematicians developed a large body of concepts that have found applications in every sphere of business and science. Because graphs are so abstract, they can represent almost anything, including an airline scheduling problem (where the vertices are cities and the edges are routes between cities), a social network (where the vertices are people and the edges show which people are friends), a computer circuit (where the vertices are logic gates and the edges are wires), or a neural network (where the vertices are neurons and the edges are chemical pathways between the neurons). Graphs can even be used to represent the abstract algebras or "groups" that physicists used to discover the fundamental particles of nature (the edges show multiplication relations between elements of the algebra). In the 1970s, computer scientists Stephen Cook, Leonid Levin and Richard Karp proved that an important class of problems in computer science—which often turn up in business when someone wants to find the most efficient way to schedule processes or to encrypt data—all reduce to the problem of colouring a graph. If someone finds a way to colour a random graph quickly, then all of the world's banking codes could be cracked.

In mathematics, abstract mental representations like graphs are powerful tools for making discoveries and solving problems. Students who develop these representations can solve a wide range of problems, which may not even appear to be related, with very little effort. For example, a simple visual tool can help a student solve the contest-level problem below.

If each column were extended, which column would the number 218 appear in?

A	B	C	D
4	6	8	10
7	10	13	16
10	14	18	22
13	18	23	28
⋮	⋮	⋮	⋮

Students typically find this problem more challenging than the letter problem I presented in chapter four (unless they have a lot of time on their hands and simply use brute force to write out the terms of every sequence until they hit on the answer). But an expert problem solver can use mental representations to solve the problem in a matter of seconds.

If you look closely at the problem, you will notice that the sequence of numbers in each column always increases by a fixed amount. These kinds of sequences, where each term is generated by adding or subtracting a fixed number to the previous term, appear everywhere in math, including in the function tables, which you may remember from high school, like the one shown here.

x	y
1	7
2	10
3	13
4	16

In middle school and high school, students usually learn to solve problems involving sequences by using formulas. But many students don't understand why the formulas work, even if they can use them correctly. They can't, for example, derive the formulas or apply them in new situations (outside of the context in which they were learned). However, when I teach students to visualize sequences on number lines, I've found they are often able to develop the formulas themselves and can even solve problems without using formulas.

Number lines are very powerful tools for solving problems because (like graphs) they are abstract. Although number lines appear everywhere in our society (for example, in the grid lines of football fields), they are cultural creations. No one has ever encountered a number line in nature. In recent studies, anthropologists have asked members of primitive tribes to place numbers on a number line and found that the tribespeople would use equal spacing for very small numbers but would crowd larger numbers more closely together. Although the problem presented above – if each column was extended, which column would the number 218 appear in? – may not appear to have any connection to number lines, students who can visualize sequences on number lines can solve problems like this one without doing any algebra.

The numbers in column A (4, 7, 10, 13 . . .) increase by 3 each time. If I add the number 3 to itself repeatedly, I generate another sequence (3, 6, 9, 12 . . .) that you should recognize if you know your times tables. These numbers, which are called the "multiples" of three, are the numbers you generate when you skip count by threes. (Note: zero is also a multiple of three, but to avoid extra verbiage, I will refer to the number 3 as "the first multiple of three," rather than using the correct phrase "the first non-zero multiple of three.")

If you plot the multiples of three and the numbers in sequence A on the same number line, as shown on page 224, you can see that the two sequences run along beside each other. The first multiple of three (3) is a fixed distance away from the first term in sequence A (it is one more), the second multiple of three is the same distance from term two of sequence A (it is also one more), and so on.

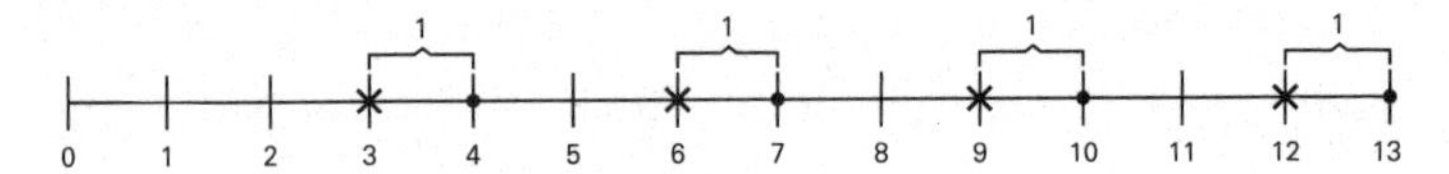

Sequence A is plotted with a dot.
The multiples of three are plotted with an X.
The terms of the sequences are always one apart.

If I ask you to find the value of the 50th term of sequence A, you might have to write out the first 50 terms of the sequence to answer my question. But if I ask you to tell me the value of the 50th multiple of three, you can find the answer by simply multiplying 3 times 50 (which is 150). The multiples of three are much easier to work with than sequence A. But because the two sequences are related, I can use my answers to questions about the multiples of three to answer questions about sequence A. Each term in sequence A is one more than the corresponding term in the sequence of multiples of three. So I know the 50th term of sequence A is just one more than the 50th multiple of three. Therefore the 50th term of sequence A is $150 + 1$, or 151. Similarly the 11th term of sequence A, for example, would be $3 \times 11 + 1 = 34$.

It's also easy to visualize division on a number line. Consider the number 13, which is in sequence A. One way to answer the question "What is 13 divided by 3?" is to ask "How many steps of size 3 do I need to take on the number line to get to 13?" I hope you can see that you need to take 4 steps of length 3 to get to 12, which is the greatest multiple of 3 before 13. But then you need to take one more step of length one to get to 13. This means that 13 divided by 3 is 4 (the four steps of length 3) with a remainder of one (the one step of length one). Using this way of visualizing division, you can see that if you divide any number in sequence A by 3 you will get a remainder of one—because all terms in sequence A are one greater than a multiple of 3. So now I know that the number 218 can't appear in column A in the problem above, because, when I divide 218 by 3, I get a remainder of 2.

I can solve the problem by checking each column in this way. The numbers in column B (6, 10, 14, 18 . . .) increase by 4 each time. If I plot the multiples of four and the numbers in series B on a number line, as shown below, I see that each number in series B is two more than the corresponding multiple of four.

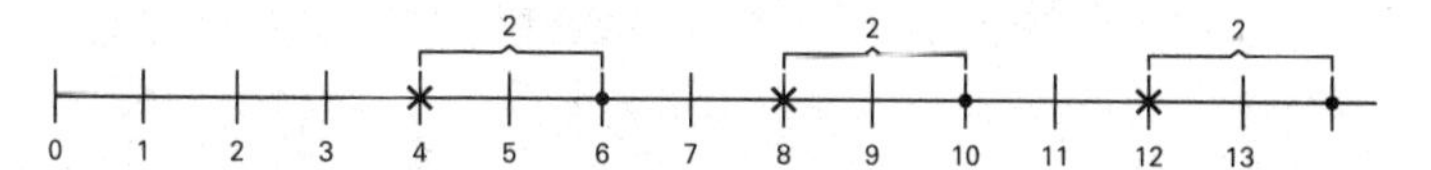

Sequence B is plotted with a dot.
The multiples of four are plotted with an X.
The terms of the sequences are always two apart.

So I know that when I divide any term in sequence B by the number 4, I will get a remainder of 2. And I also know, conversely, that if I divide a number by 4 and get a remainder of 2, the number is in sequence B. If I divide the number 218 by 4, I get a remainder of 2. So now I've solved the problem, because I know that 218 is in column B.

Students who are given time to explore sequences on number lines and guided to see relationships like these can apply the mental representations they develop to a wide range of problems. To help students develop these kinds of mental representations, JUMP has created some advanced problem-solving lessons, which you can find on the JUMP Math website. Eighty lessons for grades three to eight teach various strategies that students need to engage in high-level problem solving (or in research in math!).

To help their students become more capable and creative problem solvers, teachers at every academic level need to understand the role that abstraction plays in mathematics. Students who can use abstract mental representations—like graphs and number lines—to see the deep structure of problems in a range of fields have a huge advantage over students who never see beyond the surface details of the problems. Even younger learners can benefit from abstraction.

Some teachers are reluctant to guide the learning of younger students too much because they believe that children will naturally learn mathematical concepts on their own, by playing with "concrete materials" (blocks, toys, measuring instruments and so on). But research has shown that this view is overly simplistic. While students certainly benefit from playing with concrete objects, they usually need assistance seeing the math that is embodied in the objects. As well, concrete materials sometimes appear to impede learning. In a recent study, one group of students was instructed to use play money resembling real paper money to solve a problem while another group was given more abstract money (rectangles with numbers printed on them).[19] The group with the play money made more errors in solving the problem. Jennifer Kaminski, whose work we were introduced to in chapter three, found that grade one children learn fraction concepts more readily with grey and white circles than with pictures of objects (for instance, flowers with differently coloured petals).

These findings apply to adults as well as children. In "The Advantages of Abstract Instruction in Learning Math," Kaminski and her co-authors found that university students were more likely to correctly apply a mathematical concept in a novel situation if the concept was taught with abstract representations (letters) rather than concrete representations (measuring cups).

In spite of these results, the findings on concrete materials are mixed. While many studies have shown that concrete materials (and "perceptually rich" representations) can prevent students from generalizing concepts or transferring knowledge from one type of problem to another, other studies have found that concrete materials can help students connect math to their "real world" experiences. (In the play money study, students with the play money made more errors, but a lower proportion of their errors were conceptual.) Most researchers now recommend that teachers introduce concepts with simple concrete models or representations (avoiding materials that have too many distracting features) and that they gradually make the representations more abstract—as I did with the blocks and bags in chapter four.

Only a few of the hundreds of teachers I've met in my work with JUMP have been aware of the research on abstraction. It's hard to overstate how important it is for teachers to have access to this kind of information. Of all the factors that can be varied or controlled in the classroom (class size, use of educational technology and so on), the effect of the teacher outweighs all others. Great teachers can have an extraordinary impact on the lives of their students. For example, using school district and tax records for more than one million children, Raj Chetty, John Friedman and Jonah Rockoff studied the long-term impact of "high value added" teachers who are skilled at improving students' test scores. They found that students assigned to these teachers "are more likely to attend college, earn higher salaries, and are less likely to have children as teenagers. Replacing a teacher whose value added is in the bottom 5% with an average teacher would increase the present value of students' lifetime income by approximately $250,000 per classroom."[20]

Almost all of the teachers I know want to become better teachers and want to help their students flourish. But even the most well-motivated teacher will struggle to achieve these goals if they are required to use methods of instruction that exacerbate academic hierarchies. Rather than looking for ways to replace "low value" teachers, school districts would achieve more by providing all of their teachers with resources and professional development that are supported by rigorous evidence. If teachers were empowered to put the research on abstraction, scaffolding, variation, structure, mastery, analogy and domain-specific knowledge into practice in their classrooms, it's hard to imagine how much better their lives and the lives of their students would be. Over 50 percent of teachers in the US end up leaving their profession in the first five years, in part because the training they receive at college and in their school districts doesn't give them the tools they need to survive the challenges of the classroom. It's critical that school districts and education faculties hire coaches and instructors who know the research and who can help teachers fulfill their own potential and excel at what they do.

Creating the Conditions That Inspire Creativity

One of my high school students, Alan, was a brilliant teenager. He was also my most difficult student. After years of struggling in math, he had developed a hatred of the subject and would only do enough work to barely pass his tests. Because he was a good debater with a caustic sense of humour, he always had a clever response to my claim that his life would be better if he decided to apply his great intelligence to learning math. On one of his tests, he was asked to say in what direction and how far the graph of a parabola would shift as the numbers in its equation changed. He wrote a hilarious running commentary for each question. Beside one question he wrote:

> *This I don't know: As to whether or not the former will shift or if it will not. The latter may shift down towards the bottom of the page. That's all you could ask for I guess. People who are actually interested in this sort of thing—math I mean—would be fascinated by this illusory downward shift. However, I don't find it very absorbing.*

Beside another question he wrote:

> *The first doesn't do anything. The second is stretched by 0.1 and goes up, insofar as it can go up in the sort of insecure quasi reality that mathematicians through the ages have created for it.*

I've always had mixed feelings about my lessons with Alan. If I had been able to convince him to learn math, I expect he could have used his intelligence and creativity to do interesting work. But I was impressed by his rebellious spirit and felt the commentaries that he wrote on his tests were almost worth the aggravation he caused me as a teacher. In an ideal educational system, I would hope that students like Alan could maintain their unconventional perspective and independent spirit and still be willing to do the work that's required to develop all of their talents. (That's all you can ask for, I guess.)

In popular culture, non-conformists are often portrayed as individuals who have very different characters than ordinary people. They are rebels who are willing to take extreme risks and who don't care much about what other people think about them. But according to Adam Grant, people who do original work in art, science, politics and business are typically neither less risk-averse nor less sensitive to other people's opinions than the average person.[21] For example, comprehensive studies of entrepreneurs show that people who have little concern for pleasing others aren't more likely to become entrepreneurs, nor do their firms perform any better. And the same patterns play out in politics. Great leaders are able to challenge the status quo and instigate sweeping changes that improve the world, but these behaviours are unrelated to whether they care deeply about public approval. In most cases, exceptional leaders are pushed by the circumstances of their time to decide to become agents of change.

According to Grant, originality is not a fixed character trait—it's a free choice. His research suggests that educators and employers can create conditions in schools and workplaces that help people develop a greater sense of agency and make more ambitious and creative choices. In one experiment, Grant and a team of researchers encouraged a group of randomly selected employees from Google to view their jobs as more flexible and gave them suggestions about how they could customize their jobs to better align with their skills, interests and values. Employees in this group showed a spike in happiness and performance relative to employees who weren't encouraged to see their jobs as malleable. When the researchers added a new feature to the experiment, and encouraged employees to see both their jobs and their skills as flexible, the gains lasted six months. Employees in the experimental groups were "70 percent more likely than their peers to land a promotion or a transition to a coveted role."[22]

The research on motivation that we looked at in the last chapter suggests that schools can destroy creativity by offering extrinsic rather than intrinsic rewards for work and by making competition part of creation. Parents can also play a significant role in nurturing a spirit of independence and originality in their children. In one study, sociologists Samuel Oliner and Pearl Oliner interviewed non-Jews who risked their lives to save Jews during the Holocaust. They compared the way these individuals were raised with the way a group of neighbours who did not extend help to Jews were raised. The study revealed that "what differentiated the rescuers was how their parents disciplined bad behaviour and praised good behaviour." The parents of the rescuers were significantly more likely to use "reasoning" to change the behaviour of their children. As the Oliners discovered:

> *It is in their reliance on reasoning, explanations, suggestions of ways to remedy the harm done, persuasion, and advice that the parents of rescuers differed most. . . . Reasoning communicates a message of respect. . . . It implies that had children known better or understood more, they would not have acted in an inappropriate way. It is a mark of esteem for the listener; an indication of faith in his or her ability to comprehend, develop and improve.*[23]

In general, parents who use reasoning and who stress the importance of character and moral principles, rather than using rigid rules to guide behaviour, raise more creative children.

I've only presented a small sample of the insights that sociologists and psychologists have generated in studying creativity. This research—together with the research on deliberate practice, memory and motivation that we looked at earlier—gives me a great deal of hope for the future. Schools and businesses now have a wide range of evidence-based tools to help people learn more efficiently and think more creatively. But these tools will have limited impact on our society unless we find ways to improve the way we reason and make decisions based on evidence. Recent research in psychology and neurology has not only revealed our great intellectual potential, but it has also, unfortunately, uncovered a number of systems in the brain that cause people to frequently ignore basic principles of logic, to be swayed by stereotypes and biases, to disregard evidence, and to act against their own interests. As I will explain in the next chapter, mathematics can help us override these systems, so that we can think and act more rationally and take advantage of our great capacity to learn and innovate.

CHAPTER 8: EXTREME EQUALITY

We live in a time when even a very small change in the magnitude of a number—in the cost of borrowing money or in the concentration of an ingredient in a household product—can create a chain of effects that is felt around the planet. This was brought home recently when scientists found microplastics from cosmetics and soaps sold in North America and Europe in the bodies of Arctic fish. Because our lives are increasingly governed by numbers—and by numerical codes and algorithms that incessantly track our preferences, animate our devices and regulate our transactions—we can no longer afford to be ignorant about math.

Unfortunately our brains didn't evolve to solve the kinds of problems that our mathematical minds have created. When our prehistoric ancestors were faced with a perilous situation—for example, a ravenous dire wolf—they rarely had time to weigh all the courses of action they might take to escape the threat. So their brains developed sets of biases and heuristics (rules of thought) that were designed to limit the amount of brainpower they could dedicate to thinking about a problem before they acted. At the same time, their brains also developed a rudimentary sense of space and number that, miraculously, gave them the capacity to understand mathematics at the deepest level. (Recall from chapter three that mathematicians activate a primitive part of their brain when they do math.) This basic understanding of space and number was all their descendants needed to create technologies of enormous destructive power. And this unfortunate paradox of evolution—whereby our brains developed the capacity to understand math, but not the inclination to use mathematics (or reason) to guide our thinking—is one of the greatest impediments to our development as a species.

In this book, I have argued that people tend to significantly underestimate their true intellectual potential, particularly in mathematics. But research in psychology has revealed that we also tend to overestimate the degree to which we are rational or willing to use logic and evidence to guide our decisions. To realize our full potential as learners and thinkers, we need to understand the brain's inherent limitations as well as its strengths.

Psychologist Keith Stanovich, who has spent several decades studying the way people think, defines "dysrationalia" as the inability to think and behave rationally, despite having adequate intelligence.[1] Stanovich and other researchers have found that even highly intelligent and well-educated people often succumb to some form of dysrationalia and are susceptible to various forms of cognitive inflexibility, belief perseverance, confirmation bias (ignoring evidence that contradicts an established belief), overconfidence and insensitivity to consistency.[2]

According to Stanovich, we are all "cognitive misers" who try to avoid thinking too much.[3] Faced with problems that require careful consideration of several possible scenarios or solutions, we will often seize on an answer too quickly. Our innate mental laziness explains why, in one of Stanovich's experiments, more than 80 percent of the participants failed to answer the following question correctly:[4]

Jack is looking at Anne, but Anne is looking at George. Jack is married but George is not. Is a married person looking at an unmarried person?

Yes No Cannot be determined

Because the problem doesn't tell us whether Anne is married or not, most people immediately jump to the conclusion that the answer cannot be determined. But if you consider all of the possibilities, you will see that the answer is yes. If Anne is married, then a married person (Anne) is looking at an unmarried person (George). If Anne is unmarried, then it's still the case that a married person (Jack) is looking at an unmarried person (Anne).

The problem I've just described, determining if a married person is looking at an unmarried person, is highly artificial; it doesn't represent a situation that we would ever encounter in daily life. So the trouble that people typically have solving this problem may appear to have few practical implications. But psychologists have found that the mistakes people make when they try to solve similar problems in real life can have far-reaching consequences for the quality of our lives and the health of our society.

Traditional political and economic theories, which underpin the laws and institutions that regulate our society, assume that people are generally rational and that their thinking is normally sound. But in the 1970s, Nobel laureate Daniel Kahneman and his colleague Amos Tversky overturned this idea with a series of experiments that uncovered persistent cognitive traps that often lead people to act irrationally and against their own interests when they make decisions involving uncertainty or risk.

In *Thinking, Fast and Slow*, Kahneman formulates a wager in two different ways to illustrate one of these traps, which he calls the "framing" effect:

> *Would you accept a gamble that offers a 10% chance to win $95 and a 90% chance to lose $5?*
>
> *Would you pay $5 to participate in a lottery that offers a 10% chance to win $100 and a 90% chance to win nothing?*

According to Kahneman, people are more likely to accept the second offer—even though the offers are identical—because the thought of a *loss* evokes much stronger negative feelings than the thought of paying a *cost* that might lead to a reward.[5]

Even highly educated professionals are swayed by the way information about a risk or a reward is presented. In one study, Tversky asked physicians to say whether they would consider performing a particular type of operation that could have significant long-term benefits for patients who survived the surgery. The physicians were much more likely to choose the treatment if the risks were described in terms of survival rates (a 90 percent rate of survival) rather than mortality rates (a 10 percent rate of mortality).[6]

Kahneman and Tversky also found that people often ignore basic laws of logic or probability when they evaluate the truth of statements, especially if they are misled by their biases or by stereotypes. In one study, they asked subjects to read the following story about a fictitious person:

> *Linda is thirty-one years old, single, outspoken, and very bright. She majored in philosophy. As a student, she was deeply concerned with issues of social justice, and also participated in antinuclear demonstrations.*

After reading the story, participants were asked the following simple question:

Which alternative is more probable?
Linda is a bank teller.
Linda is a bank teller and is active in the feminist movement.

Kahneman was surprised by the number of people who thought that Linda was more likely to be a bank teller *and* a feminist. If Linda is a bank teller and a feminist, then, clearly, she must be a bank teller. But if she is a bank teller, she is not necessarily a feminist. So it's more likely that she is just a bank teller. As Kahneman recalls:

About 85% to 90% of undergraduates at several major universities chose the second option, contrary to logic. Remarkably, the sinners seemed to have no shame. When I asked my large undergraduate class in some indignation, "Do you realize that you have violated an elementary logical rule?" someone in the back row shouted, "So what?" and a graduate student who made the same error explained herself by saying, "I thought you just asked for my opinion."[7]

Following the Linda experiment, many studies have confirmed that people often have trouble assessing probability and risk, because they don't understand basic rules of probability, don't make an effort to apply those rules consistently, or are distracted by stereotypes and biases. As well, when people rate the relative importance of issues, they tend to attribute more importance to issues that are more easily retrieved from their memories. And if a person believes a statement to be true, they are also very likely to believe arguments that support the statement, even when those arguments are unsound.[8]

In the 1940s, warning of the dangers of nuclear war, Einstein wrote: "The unleashed power of technology has changed everything save our modes of thinking and thus we drift toward unparalleled catastrophe." With the acceleration of global warming, the rise of artificial intelligence, the resurgence of racism and xenophobia, and the proliferation of various forms of social manipulation and control through electronic media, Einstein's call to change our modes of thinking is more relevant now than ever.

According to the *Guardian*, firms like Cambridge Analytica (which used improperly obtained Facebook data to sway the US election in 2016) are investing heavily in "psychological operations" or "psycops" and have begun to develop effective methods of changing people's minds "through informational dominance," a set of technologies that includes rumour, disinformation and fake news.[9]

Recent political events have revealed a growing anger among the working poor and a widespread backlash against intellectual elitism that threaten to make our society less tolerant and less prosperous. I believe that much of this bitterness can be traced to a sense of frustration and learned helplessness that people begin to develop the day they enter school. Children are born with a boundless capacity for creativity, insight and wonder, but they gradually lose these qualities of mind as they learn that the various subjects their parents and teachers insisted that they study—because they were so vitally important to their future—are simply beyond their grasp. As people become less curious and less confident in their ability to understand mathematics and science, they also become less open to new ideas and more susceptible to believing false claims.

We are all influenced by the brain's unreliable heuristics, irrational biases and mental shortcuts, and we are prone to making mathematical and logical errors in everyday life. The modern state of the world only amplifies these susceptibilities and their consequences. That is why it is now particularly important to give every person the opportunity to realize their full intellectual potential. We will never move beyond the confused and corrosive debates that have become commonplace in politics and public discourse until we all learn to think more clearly and to weigh evidence more carefully. And our economies will never function properly if people don't have the conceptual tools they need to assess the real cost or value of the goods we consume and the risks involved in producing those goods.

A Simple Proposition

It's not hard to find examples of mathematical errors in the media and even in scholarly publications. For instance, in 2010, Fox News showed a pie graph divided into three parts that represented the percentage of US voters who backed various Republican candidates: 70 percent backed Sarah Palin, 63 percent backed Mike Huckabee and 60 percent backed Mitt Romney.[10] These numbers add up to more than 100 percent. Some voters may have backed more than one candidate, but then the data can't be shown on a pie graph; the segments on a pie graph show mutually exclusive possibilities.

Stories about finance or management usually involve numbers, so you would expect the writers (and editors) of business articles to understand basic math. But this isn't always the case. In 2008, the *Journal of Management Development* claimed that a firm that implemented a new way of handling customer complaints had seen a 200 percent reduction in the number of complaints.[11] This number seems a little high: once you reach a 100 percent reduction, you have no more complaints.

Although these mathematical errors are relatively innocuous, they point to a deeper problem that is rarely talked about in the media—possibly because many people in the media have low levels of numeracy. Most of the decisions we make in politics and business either explicitly or implicitly involve numbers or mathematics of some sort. It should concern us that so many adults in North America are incapable of spotting elementary mathematical mistakes and make snap decisions based on flawed heuristics and emotional triggers, especially when a nation is contemplating the economic impact of a budget, the environmental impact of a law or the social impact of a regulation. As I said in chapter two, I believe that we would have a more equitable, civil and prosperous society if everyone had a basic understanding of simple algebra, fractions, ratios, percentages, probability and statistics. I also believe that almost any adult could acquire this foundational knowledge in several weeks. It's hard to imagine how different our political debates would be if people had to demonstrate that they had this knowledge before they could vote.

Of course, knowing basic math doesn't guarantee that a person will make rational or ethical decisions. But it certainly helps. If you have learned to solve simple mathematical problems, you would instinctually consider all of the possible scenarios in Stanovich's marriage problem, because you would know that you can only find a correct solution or construct an airtight proof by keeping track of every possibility. If you can do simple calculations, you would immediately see that the two wagers described by Kahneman are identical. If you have learned to think dispassionately about problems involving probability and logic, you would be less likely to be misled by stereotypes and framing. And if you are confident in your mathematical abilities, you would embrace the mental effort required to think through the many subtle but important problems that confront our society. In my work with students of all ages, I've seen that people who experience sustained success in math usually become addicted to the thrill of meeting intellectual challenges through hard work.

One area where people struggle to make sense of numbers is in statistics. Every day we are deluged with advertisements for new drugs or diets that we are assured will produce positive results, but we don't know whether we should believe these claims. Fortunately statisticians have developed tests that can guide us in assessing a solution's effectiveness. Let's consider an example, a completely artificial one.

Suppose that (for some reason) you own a bin that contains thousands of marbles, and that 40 percent of the marbles are blue and 60 percent are red. One day you suspect that someone has stolen some blue marbles from the bin. You can't verify your suspicion directly, because there are far too many marbles to count. So you decide to take a sample. You pull 100 marbles from the bin at random and are horrified to find that only 36 of the marbles are blue. Should you count this result as evidence that someone has stolen some of the blue marbles from the bin?

Statisticians have developed a method for calculating the "*p* value" of an occurrence like this, which is the probability that the event could have happened completely by chance. In this particular case, when 40 percent of the marbles in the bin are blue and you pull out 100, it turns out that the probability that you will draw 36 or fewer blue marbles is about 25 percent. Statisticians would not consider this to be very strong evidence that some of the marbles were stolen. However, if the *p* value was less than 5 percent, they would say that the result was "significant" and that the decrease in the number of blue marbles compared with the expected value was much greater than would be expected by luck alone.

In *Knock on Wood: Luck, Chance, and the Meaning of Everything*, statistician Jeffrey Rosenthal presents a set of questions that you can ask if you want to know if an event happened by sheer luck or if you should look for a cause. According to Rosenthal, "If we can figure out which lucky events are just random, dumb luck and which are caused by actual scientific influences—which ones can be affected and which cannot—then we can make better decisions, take more reasonable actions, and better understand the world around us."[12]

To demonstrate the importance of *p* values, Rosenthal looks at the claim, made by many politicians and media pundits, that global warming is a hoax so we don't need to reduce the amount of carbon that we are pumping into the atmosphere. One of the key arguments that is made to support this claim is that even if the world has warmed in recent years, this warming is just a random statistical fluctuation that has no cause. Rosenthal did some calculations to evaluate this claim:

> *Using NASA data, I computed that the average temperature in the 37 years between 1980 and 2016 was 0.74 degrees Celsius (1.33 degrees Fahrenheit) higher than the average temperature in the 37 years between 1880 and 1916. This is quite a large difference. But is it statistically significant? Yes! The corresponding p-value (the probability that this difference would occur by luck alone) is less than one in a million billion, so it certainly isn't luck.*[13]

Rosenthal also notes that the average global temperature has increased almost one degree Celsius between 1980 and 2016, and he points out that the p value for this event is also less than one in a million billion. According to Rosenthal, this demonstrates "beyond any reasonable doubt that the yearly global temperature increases are indeed highly statistically significant and not just luck."

The vast majority of credible climate scientists believe that if our generation doesn't reduce our carbon emissions significantly, we will eventually cause many more deaths and far more destruction than all of the battles of World War II. The fact that so many voters have been convinced by non-experts that we shouldn't take any precautions to avert a potential disaster shows how ill-prepared we are to think about the most important issues of our time. In our daily lives, we know that we don't have to be certain that an accident will occur before we will take action to mitigate the risk of it happening. If 98 percent of the civil engineers in a city said that a bridge was going to collapse, no sane person would try to cross the bridge, even if 2 percent of the engineers said it was safe. And if a politician told people to put their kids in a car and drive over the bridge, because closing the bridge might force them to take a different route to work or to look for a different job or because the issue is a hoax perpetuated by other politicians, we would think they weren't fit for office.

If we understood the mathematics of risk, we would make better decisions in cases where we can't achieve certainty and where it is hard to arrive at consensus. One day, I hope we will think that politicians or pundits who make pronouncements on scientific issues that they aren't qualified to comment on—because they clearly don't understand the math—are guilty of an almost criminal level of arrogance or venality.

Mathematics is the one area of discourse where it is almost impossible to create fake news. Mathematicians occasionally make errors, but these are usually quickly spotted by their peers. Because all of mathematics can be deduced from first principles that everyone can agree on, we can be more certain about mathematical truths than about truths of any other kind. For that reason, we would be wise to build all of our beliefs and conduct all of our debates on that foundation. If every citizen had a sound knowledge of math, all of our conversations could draw from a deep well of shared beliefs and we wouldn't have to waste so much time arguing about facts or productive methods of debate.

In a society that valued and promoted intellectual equality, the average citizen would understand the importance of the scientific method and would know how to make sense of claims involving numbers. They would be able to construct valid and logically sound arguments, and would base their economic and political decisions more on data and reason and less on fake news, misleading advertising and specious arguments.

The Future of Education

I hope the examples I've provided in this book have given you confidence that it might be possible to train students to think like mathematicians by using a mathematical version of deliberate practice. But while I have argued that the training students engage in to solve chess problems is similar to the training they must engage in to solve contest problems in math, I believe that there are four features of deliberate practice in mathematics that distinguish it from the same forms of practice in most other fields. These differences make the prospect of using deliberate practice in math particularly exciting.

1. Many of the skills that people typically develop through deliberate practice (that Anders Ericsson describes in *Peak*) don't transfer very broadly. For example, people who use mnemonic techniques to remember incredibly long strings of digits don't also develop better memories. And people who are coached by a golf pro to perfect their swing don't suddenly become well-rounded athletes. In contrast, when students learn to think mathematically, they learn concepts and ways of thinking that can be applied in virtually every subject or occupation.
2. Some skills can only be developed through deliberate practice if that practice starts at an early age. For example, Ayako Sakakibara (whom we met in chapter four) found that his methods of training perfect pitch only work for students under the age of seven. But I haven't seen much evidence, in my teaching or in the research I've read in cognitive science, that there is a cut-off age for math. I didn't start my undergraduate work in math until I was thirty. And while there are some disadvantages to starting late, there are also some advantages, as adults can be more focused and disciplined than children. This means that there is hope that even our adult population could become good at math, or at least, with a relatively small effort, competent.

3. In *Peak*, Ericsson points out that deliberate practice can be tedious and gruelling and take up an enormous amount of time. When people who are engaged in deliberate practice are not getting feedback from a coach, they often spend a good deal of time honing their skills alone, and they can feel lonely and isolated. That's one of the reasons why ordinary people are unlikely to use deliberate practice to learn anything. But deliberate practice in math is best done in groups, where students can get caught up in the excitement of their peers. When students learn math in groups, they think it's fun, not gruelling. As well, people are required to learn math at school for many years. So the barriers to deliberate practice that prevent most people from becoming good chess players or golfers shouldn't stop them from becoming good at math.
4. In some fields, the body or the brain may impose strict limitations on what a person can learn through deliberate practice. If a person is short, they may never become a great basketball player (although there are some notable exceptions in the NBA). It's possible that genetic factors have some influence in mathematical achievement, but even if this is the case, I doubt the effect is very significant. The main question for educators is: how do we help people develop the motivation and

> stamina required to engage in deliberate practice in math? Einstein claimed he was successful not because he was smarter than other scientists but because he stuck with problems longer.

Many people believe that new educational technologies, such as tablet-based lessons and assessments, are our best hope for achieving significantly better outcomes in schools. I have no doubt that technology will eventually have a positive impact on education, but at present, the evidence for the effectiveness of these interventions in classrooms is mixed.[14] Before we invest too heavily in new technologies, we would be wise to conduct rigorous research to separate the fads from the genuine innovations. Otherwise we are likely to repeat the mistakes we made in the past, when we adopted expensive, well-marketed textbook programs on a massive scale, even though there was little evidence to support those adoptions.

We should be careful not to be distracted by the excitement that always surrounds new technology, in case we overlook cheaper and more practical solutions that are available now. Mary Jane Moreau produced the extraordinary results in her classroom with only pencils and paper.

Advocates of technology often argue that the new technologies will finally allow us to "customize" or "personalize" education so that students can progress at their own pace. But what if the vast majority of students could move at roughly the same pace? Could our race to personalize education prevent us from harnessing the extraordinary joy and excitement that students feel when they learn the same things at the same time? These are some of the questions we need to answer, through rigorous research, before we decide which forms of instruction are most beneficial for students.

Many people are excited about the promise of "blended learning," a computer-based approach to teaching that has been popularized by programs in which students spend a good deal of their time at school receiving tutoring on a computer (sometimes choosing their own path through the lessons). Even though these programs have not undergone much rigorous testing, they have been portrayed in the media as game-changing innovations that will transform our schools. I believe that some of the assumptions behind these approaches may be as problematic (particularly in the short term) as the ones that inspired schools to adopt discovery-based learning on a wide scale. While the assumptions appear to be progressive and student empowering, they may have downsides that advocates of blended learning programs haven't foreseen. The concerns I raise below may be relatively easy to fix, but they at least deserve to be considered.

One downside to personalized learning is that people often make poor choices when they select methods of learning or studying that they think will be effective. As psychologists Henry Roediger and Mark McDaniel explain: "Our susceptibility to illusion and misjudgment should give us all pause, especially so to the advocates of 'student directed learning.' . . .Those students who employ the least effective study strategies overestimate their learning the most and as a consequence of their misplaced confidence are not inclined to change their habits."[15] Novice learners are not only prone to select ineffective methods of learning; they are also ill-equipped to know what skills or knowledge they need to acquire to become experts in a field.

Another drawback to personalized learning is that tablet-based educational programs are not (at present) artificially intelligent. Consequently, when students can't answer questions or understand explanations presented by a computer, the teacher needs to intervene. But research has shown that many teachers aren't very good at explaining mathematical concepts, assessing student knowledge or providing remediation for students who struggle—even when all of the students in the class are working on the same topic. That's why JUMP created detailed and rigorously scaffolded lesson plans—which focus on one topic at a time—so teachers could learn the math deeply as they teach. In a classroom where students are working on many different topics at the same time and where the computers are not artificially intelligent, it is highly unlikely, even in a small class, that the average teacher will be able to intervene effectively.

In classrooms where all the students are working on different topics on computers, the teacher will also have trouble generating the group excitement or collective effervescence that I described in chapter six. This excitement, which helps students' brains work better, is critical for learning.

I believe that, in the short term, the best use of technology in the classroom is one that places the teacher (rather than the technology) at the centre of the lesson and keeps students moving forward with the same material, rather than having them all work individually on separate topics. In this model, the technology would primarily be designed to make teachers more effective, by helping them assess what their students know more quickly in the moment and by providing bonus questions that are tailored for each student (without taking students to entirely new topics or dampening the collective excitement of the class). The technology could also allow students to do more review or explore enriched topics on their own while still allowing everyone to learn the core curriculum together. The technology could even provide professional development for teachers (with interactive lesson plans, videos and coaching) and help the school system disseminate methods of teaching that are based on solid research more quickly. JUMP Math's lesson plans are available in a format that teachers can use on interactive digital whiteboards (SMART and Promethian boards), and we also plan to digitize the student resources.

Many good online, individualized tutoring programs are already helping adults learn new skills or self-motivated students learn new subjects. As personalized learning programs become more effective, they may eventually replace teachers or enhance the work of teachers in ways we can scarcely imagine now. But until these programs improve dramatically, we would be wise to test them rigorously before implementing them widely in classrooms.

Some online programs, such as Luminosity, have claimed that people can significantly improve their general cognitive functioning by playing their "brain training" games. However, in 2017 Luminosity was forced by the US Food and Drug Administration to pay a fine and retract any claims that its products help people develop transferable mental skills.[16] A recent meta-analysis found very little evidence that playing commercially available online games helps people develop cognitive abilities that transfer beyond the games themselves.[17] Primarily, this is because problem solving and conceptual work in any field requires a good deal of domain-specific knowledge and expertise.

I believe that the kind of training I received at university and through my own studies in literature, philosophy and mathematics *does* transfer broadly: it has helped me solve problems and learn new concepts quickly in many different fields. Until we find a way to enhance the performance of our brains using as yet uninvented drugs, neural implants or methods of cognitive training, it seems unlikely that any person's brain will work much more efficiently than the brain of a well-trained artist or scientist. So, for now, the best way to train students' brains may be to give them a well-rounded education, using methods of instruction that are supported by evidence. As psychologist Elizabeth Stine-Morrow has said: "My joking answer to 'Does brain training work?' is 'Yes, it's called school.'"

While new technologies will undoubtedly help us improve the state of education, teachers like Mary Jane Moreau have demonstrated that we don't have to wait for the development of the perfect computer program to instigate radical change. If we redirected the vast amounts of money that are now spent on ineffective educational resources and practices to helping teachers adopt the most effective evidence-based methods of teaching, we could produce dramatically better educational outcomes and create a more prosperous and inclusive society in a very short time, even in parts of the world that have little access to sophisticated technology.

Although I have focused on math in this book, the research on learning applies to every subject. It should be obvious that the methods of teaching I've discussed could be used to teach the sciences; the ways people think and make discoveries in science and mathematics are virtually the same. But I have also used similar methods to teach philosophy, literature and even creative writing. Finding ways to help people be more expressive and more imaginative would clearly help our society. And there are even deeper benefits that come from an engagement in the arts. The philosopher Richard Rorty said that one of the great contributions that literature and philosophy can make to our lives is to help us understand and feel what it's like to be another person. Math and science alone can't do this. Many of the great villains of history, like the Nazis, may have been good at math, but they lacked empathy and couldn't comprehend what it means to be a decent human being.

Finding Flow

A study published by the World Economic Forum in 2015 asserts that 45 percent of all jobs are in professions that run a high risk of being automated by 2025.[18] Rapid changes in the nature of work will likely produce an ever-greater concentration of wealth and eliminate many of the jobs that help give meaning to people's lives. To survive and flourish, we may need to find a sense of purpose that does not depend so much on our ability to produce physical goods or amass material wealth.

One way we might find more meaning in our lives is to draw on our ability to exist in what psychologist Mihaly Csikszentmihalyi calls a state of "flow." This means "being completely involved in an activity for its own sake. The ego falls away. Time flies. Every action, movement and thought follows inevitably from the previous one, like playing jazz. Your whole being is involved, and you're using your skills to the utmost."[19]

Great artists and scientists spend much of their lives in a state of flow. Einstein frequently described his work in euphoric or religious terms: "All religions, arts and sciences are branches of the same tree. All these aspirations are directed toward ennobling man's life, lifting it from the sphere of mere physical existence and leading the individual towards freedom."

If we could realize our full intellectual and artistic potential, our deepest sense of purpose might come from contemplating and experiencing—with the same openness to wonder we felt as children—the beauties and mysteries of existence. Our ability to worship in this way, to see and appreciate every facet of life through the lens of art and science at the same time, might be the one thing we will always do better than machines. Finding new sources of happiness and fulfillment that don't depend on pointless competition and mindless consumption might also allow us to live in a way the world can support.

When I see how quickly we are tearing apart the delicate web of life, I experience a deeper sense of pain and anxiety than most people seem to feel, perhaps because my training in mathematics has made me aware of the complexity and beauty of that web and of how hard it will be to repair when it is in tatters. In addition to filling our atmosphere and our oceans with harmful chemicals and plastics, we are increasingly running up against the constraints of the material world, as there is only so much physical space left on the Earth for us to occupy. If we all continue to insist on having a greater share of that space, and if the rich continue to amass and degrade a disproportionate amount of land, we may suffer destructive social upheavals as we damage this world irreparably.

Fortunately there is no limit to the space that we can inhabit with our minds. This vast, intangible world is full of priceless mental real estate, with edifices and structures of indescribable beauty, any one of which can be occupied by a multitude of tenants at the same time.

Rather than emigrating to Mars to avoid our most serious problems, we could find a new home here on Earth, where every member of society has the right to a rich and productive life and every economy is based on abundance and sharing rather than scarcity and greed, simply by nurturing the hidden potential that exists in all of us.

APPENDIX

HOW TO MULTIPLY COMPLEX NUMBERS

Recall, from chapter seven, that every complex number has two components. The first component of the complex number $(5, \frac{3}{4})$ is 5 and the second component is $\frac{3}{4}$.

Let (a, b) and (c, d) be a pair of complex numbers (where a, b, c and d are all real numbers). To find the product of these numbers, you must calculate the first and second components. The first component of the product is given by the expression $a \times c - b \times d$. The second component is given by $a \times d + b \times c$. Another way to express this is:

$$(a, b) \times (c, d) = (a \times c - b \times d, a \times d + b \times c)$$

For example, the product of (1, 5) and (2, 3) is given by:

$$(1, 5) \times (2, 3) = (1 \times 2 - 5 \times 3, 1 \times 3 + 5 \times 2) = (2 - 15, 3 + 10) = (-13, 13)$$

Recall that the complex number (3, 0) is equivalent to the real number 3 and the number (4, 0) is equivalent to the number 4. Using the rule for complex multiplication we can see that the product of these two numbers is (12, 0) which is equivalent to 12.

$$(3, 0) \times (4, 0) = (3 \times 4 - 0 \times 0, 3 \times 0 + 4 \times 0) = (12 - 0, 0 + 0) = (12, 0)$$

When you apply the rule for complex multiplication to a pair of real numbers a and b, the product is just the real number a × b. Or, in complex number notation:

$(a, 0) \times (b, 0) = (a \times b - 0 \times 0, a \times 0 + b \times 0) = (a \times b, 0)$

In the real number system, the number -1 doesn't have a square root, because there is no real number that will multiply by itself to give -1. However, in the complex number system, the number (-1, 0) does have a square root. If you multiply (0, 1) by itself using the rule for complex multiplication, the result is (-1, 0):

$(0, 1) \times (0, 1) = (0 \times 0 - 1 \times 1, 0 \times 1 + 1 \times 0) = (-1, 0)$

So (0, 1) is the square root of (-1, 0). In the complex number system, every number has a square root. This is one of the reasons that complex numbers are such powerful tools for problem solving: when we do algebra or manipulate equations with complex numbers, we can take the square root of any expression without worrying about whether it makes sense or not.

You can do advanced algebra that involves taking the square roots of complex numbers and you don't have to worry if the result makes sense or not. In physics, in many calculations, the second coordinate of the answer reduces to zero, so that the answer is a real number.

Chapter 1

1 Philip Ross, "The Expert Mind," *Scientific American*, August 2006.

2 "Mathematics Literacy: Proficiency Levels (2015)," Programme for International Student Assessment (PISA), National Center for Education Statistics, https://nces.ed.gov/surveys/pisa/pisa2015/pisa2015highlights_5a_1.asp.

3 Janet Steffenhagen, "Jump Math Changed My Life: Vancouver Teacher Says," *Vancouver Sun*, September 13, 2011.

4 Ibid.

5 F. W. Chu, K. vanMarle, and David C. Geary, "Early Numerical Foundations of Young Childrens' Mathematical Development," *Journal of Experimental Child Psychology* 132 (April 2015): 205–12; Greg J. Duncan et al., "School Readiness and Later Achievement," *Developmental Psychology* 43, no. 6 (November 2007): 1428–46; David C. Geary et al., "Adolescents' Functional Numeracy Is Predicted by Their School Entry Number System Knowledge," *PLoS ONE* 8, no. 1 (January 30, 2013): e5461; Melissa E. Libertus, Lisa Feigenson, and Justin Halberda, "Preschool Acuity of the Approximate Number System Correlates with Math Abilities," *Developmental Science* 14, no. 6 (August 2, 2011): 1292–1300; Michèle M. M. Mazzocco, Lisa Feigenson, and Justin Halberda, "Preschoolers' Precision of the Approximate Number

System Predicts Later School Mathematics Performance," *PLoS ONE* 6 (September 14, 2011): e23749.

6 Gavin R. Price and Daniel Ansari, "Symbol Processing in the Left Angular Gyrus: Evidence from Passive Perception of Digits," *Neuroimage* 57, no. 3 (August 1, 2011): 1205–11.

7 Roland H. Grabner et al., "Brain Correlates of Mathematical Competence in Processing Mathematical Representations," *Frontiers in Human Neuroscience* 5 (November 4, 2011): 130.

Chapter 2

1 Gerd Gigerenzer, "Smart Heuristics," in *Thinking*, ed. John Brockman (New York: HarperCollins, 2013).

2 President's Council of Advisors on Science and Technology, *Engage to Excel: Producing One Million Additional College Graduates with Degrees in Science, Technology, Engineering, and Mathematics* (Executive Office of the President, February 2012).

3 "Just the Facts: Consumer Bankruptcy Filings, 2006–2017," United States Courts, published March 7, 2018, https://www.uscourts.gov/news/2018/03/07/just-facts-consumer-bankruptcy-filings-2006-2017; "Statistics and Research," Office of the Superintendent of Bankruptcy Canada, Government of Canada, modified May 13, 2019, https://www.ic.gc.ca/eic/site/bsf-osb.nsf/eng/h_br01011.html.

4 Duncan et al., "School Readiness and Later Achievement."

5 Elisa Romano et al., "School Readiness and Later Achievement: Replication and Extension Using a Canadian National Survey," *Developmental Psychology* 46, no. 5 (September 2010): 995–1007; Linda S. Pagani et al.,

"School Readiness and Later Achievement: A French Canadian Replication and Extension," *Developmental Psychology* 46, no. 5 (September 2010): 984–94.

6 Samantha Parsons and John Bynner, *Does Numeracy Matter More?* (London: National Research and Development Centre for Adult Literacy and Numeracy, 2005).

7 Isaac M. Lipkus and Ellen Peters, "Understanding the Role of Numeracy in Health: Proposed Theoretical Framework and Practical Insights," *Health Education & Behavior* 36, no. 6 (December 2009).

8 Valerie F. Reyna et al., "How Numeracy Influences Risk Comprehension and Medical Decision Making," *Psychological Bulletin* 135, no. 6 (November 2009).

9 "Could Mental Math Boost Emotional Health?" EurekAlert! American Association for the Advancement of Science, published October 10, 2016, https://www.eurekalert.org/pub_releases/2016-10/du-cmm101016.php.

10 Steve Liesman, "'Math Has a Habit of Not Going Away'—Economists Worry Donald Trump Seems to Be Ignoring Them," CNBC, January 12, 2017.

11 Daniel J. Levitin, *A Field Guide to Lies: Critical Thinking in the Information Age* (Boston: Dutton, 2016), 9.

12 David Shenk, *The Genius in All of Us: New Insights into Genetics, Talent, and IQ* (New York: Anchor, 2011), 88.

13 Rachel Carson, *The Sense of Wonder* (Open Road Media, 2011).

Chapter 3

1 Allyson P. Mackey et al., "Differential Effects of Reasoning and Speed Training in Children," *Developmental Science* 14, no. 3 (May 2011): 582–90.

2 Shenk, *The Genius in All of Us*, 16.

3 Eleanor A. Maguire et al., "Navigation-Related Change in the Hippocampi of Taxi Drivers," *Proceedings of the National Academy of Sciences of the United States of America* 97, no. 8 (April 11, 2000): 4398–403.

4 Bogdan Draganski et al., "Neuroplasticity: Changes in Grey Matter Induced by Training," *Nature* 427, no. 6972 (January 22, 2004): 311–12; Allyson P. Mackey, Alison T. Miller Singley, and Silvia A. Bunge, "Intensive Reasoning Training Alters Patterns of Brain Connectivity at Rest," *Journal of Neuroscience* 33, no. 11 (March 13, 2013): 4796–803.

5 Carol S. Dweck, "The Secret to Raising Smart Kids," *Scientific American Mind* 18, no. 6 (December 2007): 36–43.

6 "JUMP Math in the Classroom," JUMP Math, September 28, 2016, video, https://jumpmath.org/jump/en/jump_home.

7 Marie Amalric and Stanislas Dehaene, "Origins of the Brain Networks for Advanced Mathematics in Expert Mathematicians," *Proceedings of the National Academy of Sciences of the United States of America* 113, no. 18 (May 3, 2016): 4909–17.

8 Jordana Cepelewicz, "How Does a Mathematician's Brain Differ from That of a Mere Mortal?" *Scientific American*, April 12, 2016.

9 Jennifer A. Kaminski and Vladimir M. Sloutsky, "Extraneous Perceptual Information Interferes with Children's Acquisition of Mathematical Knowledge," *Journal of Educational Psychology* 105, no. 2 (May 2013): 351–63.

10 David H. Uttal et al., "The Malleability of Spatial Skills: A Meta-analysis of Training Studies," *Psychological Bulletin* 139, no. 2 (March 2013): 352–402.

11 Shenk, *The Genius in All of Us*, 29.

12 The term "structured inquiry" was suggested to me by Brent Davis, who is a Distinguished Research Chair in Mathematics Education at the University of Calgary. I discuss his work in chapter five.

Chapter 4

1 Anders Ericsson and Robert Pool, *Peak: How to Master Almost Anything* (New York: Viking, 2016), xiv.

2 Daniel T. Willingham, *Why Don't Students Like School? A Cognitive Scientist Answers Questions about How the Mind Works and What It Means for the Classroom* (San Francisco: Jossey-Bass, 2009), 3.

3 Ibid., 133.

4 Ibid., 3.

5 Amy Bastian, "Children's Brains Are Different," in *Think Tank: Forty Neuroscientists Explore the Biological Roots of Human Experience*, ed. David J. Linden (London: Yale University Press, 2018) excerpted in *Johns Hopkins Magazine* (Summer 2018), https://hub.jhu.edu/magazine/2018/summer/human-brain-science-essays/.

6 Bastian, "Children's Brains Are Different."

7 @OctopusCaveman, Twitter, August 26, 2018, 7:56 a.m., https://twitter.com/octopuscaveman/status/1033578911697784832, included in "18 Parent Tweets That Basically Sum Up Having Kids," *BrightSide*, September 16, 2018.

8 Ericsson and Pool, *Peak*, 172.

9 Ashutosh Jogalekar, "Richard Feynman's Sister Joan's Advice to Him: 'Imagine You're a Student Again,'" *The Curious Wavefunction*, April 2, 2017, http://wavefunction.fieldofscience.com/2017/04/richard-feynmans-sister-joans-advice-to.html.

10 Adam Grant, *Originals: How Non-conformists Move the World* (New York: Penguin Books, 2016), 9.

Chapter 5

1 "Our Story," ResearchED, https://researched.org.uk/about/our-story/.

2 Barak Rosenshine, "Principles of Instruction: Research-Based Strategies That All Teachers Should Know," *American Educator* 36, no. 1 (Spring 2012): 12.

3 E. D. Hirsch Jr., *Why Knowledge Matters: Rescuing Our Children from Failed Educational Theories* (Cambridge, MA: Harvard Education Press Group, 2016), 88.

4 Daniel T. Willingham, *The Reading Mind: A Cognitive Approach to Understanding How the Mind Reads* (San Francisco: Jossey-Bass, 2017), 110.

5 Hirsch, *Why Knowledge Matters*, 89.

6 K. Anders Ericsson, "An Introduction to *The Cambridge Handbook of Expertise and Expert Performance*: Its Development, Organization, and Content," in *The Cambridge Handbook of Expertise and Expert Performance*, ed. K. Anders Ericsson (Cambridge: Cambridge University Press, 2012), 13.

7 John R. Anderson, Lynne M. Reder, and Herbert A. Simon, "Applications and Misapplications of Cognitive Psychology to Mathematics Education," *Texas Education Review* (Summer 2000): 13.

8 John Dunlosky et al., "The Science of Better Learning: What Works, What Doesn't," *Scientific American Mind* (September 2013): 43.

9 D. Rohrer and K. Taylor, "The Shuffling of Mathematics Problems Improves Learning," *Instructional Science* 35, no. 6 (2007): 481–98.

10 Paul A. Kirschner, John Sweller, and Richard E. Clark, "Why Minimal Guidance during Instruction Does Not Work: An Analysis of the Failure of Constructivist, Discovery, Problem-Based, Experiential, and Inquiry-Based Teaching," *Educational Psychologist* 41, no. 2 (June 2006): 75–86.

11 Louis Alfieri et al., "Does Discovery-Based Instruction Enhance Learning?" *Journal of Educational Psychology* 103, no. 1 (February 2011): 1–18.

12 Armando Paulino Preciado-Babb, Martina Metz, and Brent Davis, "The RaPID Approach for Teaching Mathematics: An Effective, Evidence-Based Model," University of Calgary Paper presented at the Canadian Society for the Study of Education Annual Conference, University of British Columbia, Vancouver, BC (June 1–5, 2019). (The paper is available on ResearchGate.) The research partnership between JUMP Math and the University of Calgary is called "Math Minds." According to the researchers, good teachers (or good resources) "unravel" concepts into smaller conceptual threads and then help students notice connections between the threads and concepts they have learned previously. They also help students weave the threads into coherent conceptual wholes. For example, in the long division lesson described in chapter 4, I help students notice the connection between the numbers in the

long division algorithm and the numbers in their diagrams and the connection between the "bring down" step and the regrouping of the pennies. In this book I have used the word "steps" when I describe how I break concepts into more manageable chunks for students (because the word is familiar), but the researchers avoid this term, as they feel it doesn't fully describe the process of teaching, or even how JUMP lessons are structured. They prefer the metaphor of conceptual threads.

13 Benjamin S. Bloom, "The 2 Sigma Problem: The Search for Methods of Group Instruction as Effective as One-to-One Tutoring," *Educational Researcher* 13, no. 6 (June 1984): 4–16.

14 Thomas R. Guskey, "Lessons of Mastery Learning," *Educational Leadership: Interventions That Work* 68, no. 2 (October 2010): 52–57; Stephen A. Anderson, "Synthesis of Research on Mastery Learning," ERIC Document Reproduction Service No. ED 382567 (November 1, 1994); Thomas R. Guskey and Therese D. Pigott, "Research on Group-Based Mastery Learning Programs: A Meta-analysis," *Journal of Educational Research* 81, no. 4 (March 1988): 197–216; Chen-Lin C. Kulik, James A. Kulik, and Robert L. Bangert-Drowns, "Effectiveness of Mastery Learning Programs: A Meta-analysis," *Review of Educational Research* 60, no. 2 (June 1, 1990): 265–99.

15 Kate Wong, "Jane of the Jungle," *Scientific American* 303, no. 6 (December 2010): 86–87.

16 https://www.poemhunter.com/poem/to-posterity/ (trans. H. R. Hays).

Chapter 6

1 Peter C. Brown, Henry L. Roediger III, and Mark A. McDaniel, *Make It Stick: The Science of Successful Learning* (Cambridge, MA: Harvard University Press, 2014), 145–46.

2 Willingham, *Why Don't Students Like School*, 56.

3 Ibid., 163.

4 Daniel Pink, *Drive: The Surprising Truth about What Motivates Us* (New York: Riverhead Books, 2009), 7.

5 Ibid., 8.

6 Deborah Stipek, "Success in School—for a Head Start in Life," in *Developmental Psychopathology: Perspectives on Adjustment, Risk, and Disorder*, ed. Suniya S. Luthar et al. (Cambridge: Cambridge University Press, 1997), 80.

7 Sian L. Beilock et al., "Female Teachers' Math Anxiety Affects Girls' Math Achievements," *Proceedings of the National Academy of Sciences of the United States of America* 107, no. 5 (February 2, 2010): 1860–63.

Chapter 7

1 Friedrich Nietzsche, *Menschliches, Allzumenschliches (Human, All-Too-Human)*, 1878, cited in Shenk, *The Genius in All of Us*, 48.

2 Beethoven quoted in Shenk, *The Genius in All of Us*, 48.

3 Dean Simonton, "Your Inner Genius," *Scientific American Mind* 23, no. 4 (Winter 2015): 7.

4 Grant, *Originals*, 172.

5 Leonardo da Vinci's diaries cited in Michael J. Gelb, *How to Think Like Leonardo da Vinci: Seven Steps to Genius Every Day* (New York: Delacorte Press, 1998), 50.

6 Todd Kashdan et al., “The Five Dimensions of Curiosity,” *Harvard Business Review* (September–October 2018): 59–60.

7 Ibid., 59.

8 Claudio Fernández-Aráoz, Andrew Roscoe, and Kentaro Aramaki, “From Curious to Competent,” *Harvard Business Review* (September–October 2018).

9 Frank Dumont, *A History of Personality Psychology: Theory, Science, and Research from Hellenism to the Twenty-First Century* (Cambridge: Cambridge University Press, 2010): 474.

10 Celeste Kidd and Benjamin Y. Hayden, “The Psychology and Neuroscience of Curiosity,” *Neuron* 88, no. 3 (November 4, 2015): 449–60.

11 E. Marti-Bromberg et al., “Midbrain Dopamine Neurons Signal Preference for Advance Information about Upcoming Rewards,” *Neuron* 63, no. 1 (July 2009): 119–26.

12 Lewis Campbell and William Garnett, *The Life of James Clerk Maxwell* (London: Macmillan, 1882).

13 “Keep It Simple?,” editorial, *Nature Physics* 7 (June 1, 2011), https://doi.org/10.1038/nphys2024.

14 Mary L. Gick and Keith J. Holyoak, “Analogical Problem Solving,” *Cognitive Psychology* 12 (1980): 351.

15 Dedre Gentner and Jeffrey Lowenstein, “Learning: Analogical Reasoning,” in *Encyclopedia of Education*, 2nd ed., ed. James W. Guthrie (New York: Macmillan, 2003).

16 Dedre Gentner, “Structure-Mapping: A Theoretical Framework for Analogy,” *Cognitive Science* 7, no. 2 (April 1983): 155–70.

17 M. Vendetti et al., “Analogical Reasoning in the Classroom: Insights from Cognitive Science,” *Mind, Brain and Education* 9, no. 2 (June 2015): 100–106, block quote from

p. 103 references Lindsey E. Richland and Ian M. McDonough, "Learning by Analogy: Discriminating between Potential Analogs," *Contemporary Educational Psychology* 35, no. 1 (January 2010): 28–43.

18 Vendetti et al., "Analogical Reasoning in the Classroom." See also: Dedre Gentner, Nina Simms, and Stephen Flusberg, "Relational Language Helps Children Reason Analogically," in *Proceedings of the 31st Annual Conference of the Cognitive Science Society*, ed. Niels A. Taatgen and Hedderick van Rijn (Cognitive Science Society, 2009), 1054–59; Benjamin D. Jee et al., "Finding Faults: Analogical Comparison Supports Spatial Concept Learning in Geoscience," *Cognitive Processing* 14, no. 2 (May 2013): 175–87; Bryan J. Matlen, "Comparison-Based Learning in Science Education," unpublished doctoral dissertation (Carnegie Mellon University, 2013); Norma Ming, "Analogies vs. Contrasts: A Comparison of Their Learning Benefits," in *Proceedings of the Second International Conference on Analogy*, ed. Boicho Kokinov, Keith Holyoak, and Dedre Gentner (Sofia, Bulgaria: New Bulgarian University Press, 2009), 338–47; Linsey Smith et al., "Mechanisms of Spatial Learning: Teaching Children Geometric Categories," *Spatial Cognition* 9 (2014): 325–37.

19 Nicole M. McNeil, David H. Utta, Linda Jarvin, and Robert J. Sternberg, "Should You Show Me the Money? Concrete Objects Both Hurt and Help Performance on Mathematics Problems," *Learning and Instruction* 19 (2009): 171–84.

20 Raj Chetty, John N. Friedman, and Jonah E. Rockoff, "Measuring the Impact of Teachers II: Teacher

Value-Added and Students' Outcomes in Adulthood," *American Economic Review* 104, no. 9 (2014): 2633–79.

21 Grant, *Originals*, 23.

22 Ibid., 24.

23 Ibid., 163–64.

Chapter 8

1 Kurt Kleiner, "Why Smart People Do Stupid Things," *University of Toronto Magazine* (Summer 2009): 36.

2 Keith E. Stanovich, "The Comprehensive Assessment of Rational Thinking," *Educational Psychologist* 51, no. 1 (2016): 1–10.

3 Ibid., 7.

4 Kleiner, "Why Smart People Do Stupid Things," 36.

5 Daniel Kahneman, *Thinking, Fast and Slow* (Toronto: Anchor Canada, 2011), 364.

6 Ibid., 367.

7 Ibid., 158.

8 Ibid., 8.

9 Carole Cadwalladr, "'I Made Steve Bannon's Psychological Warfare Tool': Meet the Data War Whistleblower," *Guardian*, March 18, 2018.

10 Levitin, *A Field Guide to Lies*, 10.

11 Ibid., 6.

12 Jeffrey S. Rosenthal, *Knock on Wood: Luck, Chance, and the Meaning of Everything* (New York: HarperCollins, 2018), 13.

13 Ibid., 126.

14 For example, in an article in the *Hechinger Report* in July 2019, computer scientist and educator Neil Heffernan says: ". . . studies have shown that student-paced learning tools

may sometimes exacerbate achievement gaps. A 2013 meta-analysis by Duke University researchers of 23 studies examining the efficacy of 'intelligent' tutoring systems showed that self-paced education technology that personalizes learning for each student worsens achievement gaps by allowing already highly motivated students to progress while leaving unmotivated students in the dust. On the other hand, this same meta-analysis showed that systems that were part of a teacher-led homework routine did not worsen achievement gaps and led to increased student learning. Nightly online homework, monitored by a teacher, may help to close achievement gaps."

15 Brown, Roediger, and McDaniel, *Make It Stick*, 123–24.

16 Ed Yong, "The Weak Evidence behind Brain-Training Games," *Atlantic*, October 3, 2016, https://www.theatlantic.com/science/archive/2016/10/the-weak-evidence-behind-brain-training-games/502559/.

17 Ibid.

18 Kathleen Elkins, "The Radical Solution to Robots Taking Our Jobs," World Economic Forum, June 9, 2015, https://www.weforum.org/agenda/2015/06/the-radical-solution-to-robots-taking-our-jobs/.

19 John Geirland, "Go with the Flow," *Wired*, September 1, 1996.

PERMISSIONS

Quotations on pages 105 and 106 are from "Children's Brains Are Different" by Amy Bastian in *Think Tank: Forty Neuroscientists Explore the Biological Roots of Human Experience.* Published by Yale University Press. Reprinted by permission.

Quotations on pages 166 and 254 are from *Make It Stick: The Science of Successful Learning* by Peter C. Brown, Henry L. Roediger III and Mark A. McDaniel. Published by Belknap Press (an imprint of Harvard University Press). Reprinted by permission.

Quotation on pages 89–90 is from *Peak: How to Master Almost Anything* by Anders Ericsson and Robert Pool. Published by Viking Books. Reprinted by permission.

Quotation on page 259 is from "Go with the Flow" by John Geirland. Published in *Wired.* Reprinted with permission from Condé Nast.

Quotation on page 211–12 is from "Analogical Problem Solving" by Mary L. Gick and Keith J. Holyoak. Published in *Cognitive Psychology.* Reprinted by permission.

Quotation on page 130 is from *Why Knowledge Matters: Rescuing Our Children from Failed Educational Theories* by E. D. Hirsch Jr. Published by Harvard Education Press Group. Reprinted by permission.

Quotation on page 127 is from "Principles of Instruction: Research-Based Strategies That All Teachers Should Know" by Barak Rosenshine. Published in *American Educator*. Reprinted by permission.

Quotations on pages 246 and 247 are from *Knock on Wood: Luck, Chance, and the Meaning of Everything* by Jeffrey S. Rosenthal. Published by HarperCollins. Reprinted by permission.

Quotation on pages 66–67 is from *The Genius in All of Us: New Insights into Genetics, Talent, and IQ* by David Shenk. Published by Anchor Books. Reprinted by permission.

Quotations on pages 94, 95–96, 96–97 and 167 are from *Why Don't Students Like School: A Cognitive Scientist Answers Questions about How the Mind Works and What It Means for the Classroom* by Daniel Willingham. Published by Jossey-Bass. Reprinted by permission.

Every effort has been made to contact the copyright holders; in the event of an inadvertent omission or error, please notify the publisher.

INDEX

JOHN MIGHTON is a mathematician, an author and the founder of JUMP Math, a charity dedicated to helping people fulfill their potential in math. His first two books, *The Myth of Ability* and *The End of Ignorance*, were national bestsellers. Mighton has been recognized as an Ashoka Fellow, awarded five honorary doctorates for his lifetime achievements and appointed as an Officer of the Order of Canada. He is also an award-winning playwright: he's received the prestigious Siminovitch Prize in Theatre, two Governor General's Literary Awards for Drama, the Dora Award and the Chalmers Award. He lives in Toronto.